The Grace of Dogs

THE
GRACE
OF
DOGS

A Boy, a Black Lab, and

a Father's Search for the

Canine Soul

Andrew Root

Convergent Books
New York

Copyright © 2017 by Andrew Root

All rights reserved.
Published in the United States by Convergent Books, an imprint of the
Crown Publishing Group, a division of Penguin Random House LLC,
New York.
convergentbooks.com

CONVERGENT BOOKS is a registered trademark and its C colophon
is a trademark of Penguin Random House LLC.

Page 5 photograph © 2017 by Andrew Root; page 9, © 2017 by seh342;
page 141, © 2017 by brackish_nz.

Library of Congress Cataloging-in-Publication Data is available upon
request.

ISBN 978-0-451-49759-8
Ebook ISBN 978-0-451-49760-4

Printed in the United States of America

Jacket design by Jessie Sayward Bright
Jacket photographs by debibishop (dog); Chris Oberthaler/EyeEm (wood)

10 9 8 7 6 5 4 3 2 1

First Edition

To Wally Ekstrand (1925–2000),

Grandpa, who always loved me

and whom dogs always loved

CONTENTS

Kirby Leaves

One afternoon in late July 2013, just weeks after his eleventh birthday, our black Lab, Kirby, wouldn't move. All afternoon, he lay at the foot of the stairs with a pained, heavy look in his eye. That night, for the first time in family memory, he failed to make it to my son, Owen's, room to sleep beside him. Instead, Kirby stayed on the cool bathroom floor, a place he rarely went, let alone slept.

Kirby's intermittent bad days had started nearly a year earlier, when he struggled to make it up the stairs or couldn't summon the energy to chase tennis balls. Yet, every time before, after a day or two of exhaustion, he'd always rally enough to resume playing in the yard with our kids, then eight-year-old Owen and five-year-old Maisy. So, on this day in July, confident that Kirby's illness was temporary, my wife, Kara, decided to take him to the vet for his next exam.

Something was wrong, though. Kara labored mightily to get our slow, reluctant Lab into and out of the car. Kirby, never the kind of dog to voice displeasure, growled, groaned, and pulled on the leash before finally consenting.

At the vet's office, Kara heard the words we had dreaded ever since we first fell in love with Kirby's floppy black ears. The vet had found a large mass in Kirby's stomach; our dog was in terrible pain, and the end was here. The vet said he shouldn't even be moved again. Kara's anguish bled into her voice when she called to tell me. She was coming to pick up me and the kids so we could all be with Kirby one last time.

When the four of us arrived back at the vet's office, Kirby was lying inert on the sterile linoleum floor, his chest moving in ragged bursts. Each shallow breath was work. Owen and Maisy threw themselves onto him, wailing. Kirby mustered just enough energy to raise his chin and gently lick Maisy's nose. Owen hugged Kirby's neck, screaming his grief like a mother who had just lost her son—"No! No! No!"

The vet entered and knelt next to Kirby, holding a syringe loaded with a medicine that would take away his pain but also his life. Owen stayed put at Kirby's side; he refused to allow his friend to depart alone. As the vet gently inserted the needle into a spot she had shaved on Kirby's back leg, Owen announced to the room, and perhaps to the universe, "My face will be the last thing Kirby sees."

Owen rested his nose against Kirby's, locking eyes, and I watched my son as the light in his dog's eyes went dim. All the while, Owen kept his arms around his pal's head, his tears wetting the muzzle of the dead dog. I couldn't take it. I took Maisy by the hand and left the room. I had known sadness would come, but I was surprised to feel a rush of anger at the thought that Kirby would never return. I headed outdoors with my daughter to feel the grass under my feet.

I'll never forget Kirby's death, but what I remember most about that day is what happened afterward, in that same room, between the boy and his departed dog. When Maisy and I came back inside, Kara was sitting with Owen while he petted and embraced Kirby and continued to cry. Owen knew that his best friend was gone, but he wasn't ready to say good-bye. I watched as he quieted, stood, wiped his cheeks, and said to his mom, "I will be right back."

Owen walked out to the lobby and returned with a dog treat and a paper cup he'd filled with water. Silently and purposefully, he knelt before Kirby's body, placed the tiny dog bone on Kirby's back, and, dipping his finger in the water, reverently made the sign of the cross on Kirby's forehead. Then he lifted his hands to heaven like a priest at the altar, looked up, and whispered, "I love you, Kirby. Good-bye."

That's the image I can't shake.

The Origins of Kirby

Kirby is, to date, the most outrageous impulse buy of my life. I'm not tempted by shiny new gadgets or even those candy bars that line the checkout counter at the grocery store. But this was different.

It was the summer of 2002, and Kara and I had recently moved from Los Angeles to Princeton, New Jersey. We'd left behind the jammed freeways and entertainment industry vibe of LA for Princeton's Colonial buildings and dense air of academic importance. I had survived my introduction to Princeton Seminary's PhD program, slogging through a grueling summer session of German.

Both Kara and I grew up in suburbs of the Twin Cities, thirty minutes away from each other, but we met in grad school in Southern California. It was the late nineties. I wore mainly windpants and backward baseball caps and spent my spare time watching college hockey.

Kara, with her long, dark, curly hair and combat boots, was sometimes mistaken for Alanis Morissette, and she preferred run-down coffee shops to ESPN. After months of being "safe" friends with nothing in common, a summer of Intensive Biblical Greek and awkward study sessions (that is, make-out sessions) led to us dating. A few months later, we were engaged.

Both of us had grown up with hunting dogs. Kara's was a black Lab named Mitzie who actually hunted, while mine, a Brittany spaniel named Katie, lay on the bed all day and would run away when anyone opened the front door. As our marriage began, in student housing and tiny Pasadena apartments, having a pet was not on our minds. Our life was transitional, and we never knew what would come next.

As the nineties gave way to a new century, clarity began to emerge in the form of acceptance and rejection letters, and soon we were packing a U-Haul for the East Coast. In retrospect, we'd always kind of known we would get a dog; we just didn't know when.

Or that it would arrive with ice cream.

One Friday evening, just before the start of fall term, we found ourselves in need of milk. So we headed off to Halo Farm, just outside Trenton, which we'd discovered had cheap, fresh milk and the bonus of amazing homemade ice cream. In the short time we'd lived in Princeton, the weekly trip to Halo Farm had become part of our routine. And on every trip, we'd noticed the sign by the side of the road: PetWorld.

The last thing we needed during this season of our lives was a furry animal taking up room in our one-bedroom apartment. For some reason, though, this time around the pull was too strong. On the way home, with the floor of the car filled with cartons of milk and double chocolate ice cream, we pulled in.

"Let's just pop in for a look," we said, "just to see the puppies." Maybe we'd pet one or two. Maybe we'd ask the attendant about Labradors—we'd always wanted a Labrador—but only to get some information for the future. After all, we weren't the kind of people who'd buy a dog at a pet store. We would do our homework and support a local breeder, and only when the time was right.

At the back of the store was a cage holding a small black puppy. We peered inside at the naked pink belly of a sleeping little Lab who was panting like a fat man on a long run. I've seen dogs "hunt" in their sleep before—the quivering legs, the whimpering, the occasional sleep bark—but from the looks of it, this tiny guy was working an entire field filled with tennis balls or squirrels or whatever prey he had in his seven-week-old imagination. He looked exhausted, and he hadn't even woken up!

As we stood watching his tiny chest rise and fall with the rhythm of a bouncing basketball, the clerk asked if we wanted to hold him. The correct answer would have been "No, thank you. We have milk and double chocolate ice cream in the car, and we need to get home." Yet somewhere between my brain and mouth, those words turned into "That'd be great!"

Fifteen minutes later, the back of our little Honda Civic was packed with bags of dog food, a kennel, and toys. An eight-pound puppy curled up nervously on Kara's lap. As if we needed any more proof that this was a major impulse buy, the ice cream sitting on the floor of the car hadn't even begun to melt.

We debated long and hard what to name him, this overwhelming, excitable black Lab who had burrowed so suddenly into our home and hearts. The other theology nerds at Princeton had dogs named after theologians. There was a Scottie named Schleiermacher, a spaniel mix named Augustine, and a handful of "Calvins" of all shapes and sizes. So I decided to buck the trend and return to my youth for a name: Kirby Puckett, the Minnesota Twins centerfielder who won six Gold Gloves and two World Championships. For me, as a Minnesota boy who grew up in the eighties and early nineties, there was no name more revered than Kirby. It was perfect.

From the beginning, Kirby the dog was a ball-catching sensation, just like his namesake. He became one of the fastest fetchers in the neighborhood, known for being able to catch a tennis ball in his mouth, no matter how hard it was thrown, without flinching. In the winter, little kids would line up to throw snowballs at his face, knowing he'd snatch every one of them. Kirby spent most of his life carrying around a yellow tennis ball, shifting it from one jowl to the other, leaving his big pink tongue hanging in the fresh air.

After only a few short months, it was hard to imagine

there had ever been a Kara and me without a Kirby. He had become part of us, and our lives took on a whole new shape with the added joy of his participation. It's strange to say this now as a parent of two kids, but in a very real sense, it was Kirby who first turned us into a family.

A Mind of His Own

When Kirby was eight months old, I decided it was time to start taking him with me on runs. I had always wanted a dog I could jog with, and Kirby was now big and strong enough to be my partner in exercise. Since he could never get enough of playing outside, I figured he would enjoy working out his puppy energy by galloping for miles alongside his owner.

So one morning I laced up my running shoes, grabbed the leash, and said, "Kirby, you want to go for a run?" Naturally the answer was yes. We took off in our usual direction for walks, toward the field where we always played fetch. Kirby ran happily beside me. It was just as I'd always imagined: me jogging with my loyal dog at my heel, his quiet presence and companionship fueling me as the blocks turned into miles.

It didn't last long. Within yards of passing the field where we normally stopped to play, Kirby threw on the brakes. The leash jerked in my hands, and I turned to see him sitting down, looking over his shoulder at the field. I could see what he was thinking, "What is *this*? I don't run for running's sake. Are you crazy? I run only if there's a *ball* involved!"

I pleaded with Kirby to continue, but he would have none of it. He plopped down flat on the sidewalk, the better to anchor himself, again looking over his shoulder at the field and then back at me. In the end, he won. My only recourse was to go back home.

The next morning, I had a new plan. I convinced Kara to run with me, figuring that if she ran just a few feet ahead of us, Kirby would see our jogging as a game of chase and go the couple miles I wanted him to. But again, he was having none of it. A block or two in, he lay down just like before.

The next morning, we tried again, this time with treats! No luck.

The following morning, we did it with Kara holding a tennis ball. Nope.

No matter what we tried, he'd get as far as the field, lie down, and refuse to budge. We even tried carrying him past the field, thinking once he'd passed it he'd forget about play and join the *super fun* jog! Kirby wasn't fooled for a second.

It was absurd, but Kirby didn't care. He had his own opinions and his own intentions. Just because I wanted my dog to jog didn't mean my dog was going to jog. Kirby wasn't being lazy. He was telling me that if I wanted to be a pet owner, I'd have to accept that he was his own man.

A Kid's Best Friend

Kirby was two years old when Owen was born. Hours after our son pushed his way into the world, filling our hearts to overflowing with love, I returned to our apartment. I felt this odd need to tell Kirby what had happened, to pull him into this moment of wonder. He was our family, after all.

Kara and I had read in some baby book that before bringing the baby home from the hospital, dog owners should introduce their dog to an article of the new baby's clothing; the dog will sniff the scent of the newborn and prepare for its entrance into his space. As an obedient new father, I took home the little hat Owen wore in the first hours of his life and held it next to Kirby's nose. Yet there was no epiphany. He stopped and sniffed for a good twenty seconds, then raced off to find a tennis ball for me to throw. There was no sense that he understood the amazing event that had just happened to us, and I knew no other way to communicate it to him.

Even when Owen came home, Kirby seemed to pay him no mind. There were a few moments when he got a little insistent that we put down this smelly, noisy thing that was taking up our attention and throw him the darn tennis ball already. For the most part, though, Kirby seemed to treat Owen like another piece of furniture.

All that changed when we gave squirming Owen a bath after dinner. All of a sudden, Kirby was transfixed. He sat next to the tub, tennis ball in mouth of course, observing Owen and trying to make sense of why this little piece of furniture was now wiggling. Then there was the water. Kirby liked to drop his tennis ball in the tub, bob it with his paw, then snatch it out like a freshly caught fish, ignoring the baby who shared this particular pond.

One night, Kara carried Owen from the tub and laid him, still naked, on a blanket on the living room floor. Kirby was in the corner pushing his tennis ball under a shelf in a solo game of hide-and-seek. As Owen lay kicking on the blanket, Kirby approached him slowly. It was the first direct encounter Kirby had had with Owen's infant bottom without the huge, scented disposable diaper getting in the way. Kirby slowly advanced toward the baby, right front paw raised in investigation. He began sniffing curiously, inching his nose nearer to Owen's freshly bathed baby butt. Suddenly, his paw dropped, his ears went back, his tight body loosened, and he raced to the corner, returning with a tennis ball. He dropped the ball next to Owen, pushing it toward his tiny face, and then lay down next to Owen, nudging his arm and licking his cheek. Owen squealed with delight and flailed his arms and legs.

From that night on, Kirby and Owen were best friends. Somehow, Kirby had come to know that Owen was a person—a ball-throwing, fur-petting human person like Kara and me. And he knew reflexively what to do for this new person: be near him, watch over him, play with him. A night didn't go by without Kirby sleeping next to Owen's bed, watching over him, giving the little boy his presence as a gift. Kirby became Owen's ball-catching entertainment, his pillow during midday episodes of *Sesame Street*, and his daily snack partner, settling in with his bowl of Cheerios or Goldfish for their routine, "One for me! And one for DeeDee!" (Owen's name then for Kirby).

Two and a half years later, we brought home Owen's little sister, Maisy. Kirby was five by then, and already experienced with a little kid, so we knew what to do with the introduction. During the first diaper change, we let him sniff Maisy's butt, providing the information that she, like Owen, was also a human being. Kirby got the message and seemed to love Maisy as he did Owen, lying next to her, watching over her. He had already committed to sleeping every night with Owen, but he made up the difference by lying next to Maisy every morning, bathing her giggling face in kisses.

Like her brother, Maisy loved Kirby back. She loved his wet tongue on her face, and as she grew, she loved throwing her body onto Kirby's as the dog lay on the floor. Whenever Kirby entered the room, little Maisy would shriek with excitement.

As Maisy progressed from grunts and gleeful shrieks to something like language, she would shout, "Daaga, daaga!" at Kirby. Wanting him to sit near her, she'd point and command, "Daaga, daaga, come!" And remarkably, Kirby did. He wanted to be with Maisy, too; wanted to be her friend.

Truth be told, despite all the time Kara and I had spent with him, Kirby became more our kids' dog than ours. When Maisy started kindergarten, Kirby would lie next to her in the morning as she fretted about the possibility of having a bad day. He would celebrate Owen's every return home with tail wagging and eyes gleaming with joy. At Halloween, Kirby played a dignified Superman with a red cape tied around his neck. On hot summer afternoons, when our backyard filled with children, Kirby planted himself in the middle of things, sharing the kiddie pool, never wanting to leave. Believing himself to be a seven-year-old, he became the center of their games, the lead character in their spectacular zoo, the wise creature in their mysterious jungle.

Losing a Friend

It was about five years after Maisy was born that the limp set in and gray appeared on Kirby's chin. I'd always known our dog wouldn't be with us forever, but now that fact was becoming real. I knew that when Kirby left us it would break Owen and Maisy. I knew it would break me as well.

When Kirby did die, we were astonished at the number of Facebook condolences, and even cards in the mail, that we received. People came out of the woodwork, sharing how hard the loss of their dogs had been, treating us with so much gentleness that it was as though a human family member had passed away. These were the messages of people who had loved and lost, who had lost a dear partner in a bond of love.

Then there was Owen's liturgy on that veterinary clinic floor, the moment that had so surprised me, and that had seemed so beautifully *right*. The holy, lingering shadow of my son's rite at Kirby's death, the memories of the bond Kirby had had with us all, the grief that engulfed our house, and the words of sympathy and solidarity from others that flooded in over the next days and weeks—I found myself endlessly mulling over what it all meant. What was this thing we'd experienced with this animal? Why did the loss of this being hurt so bad? Why did Owen's sacramental act feel so appropriate? The questions wouldn't leave me.

Lately, scientists and researchers have been popping up all over talk radio and twenty-four-hour news, lauding amazing discoveries and scientific breakthroughs in our understanding of dogs—who, for most of scientific history, have generally been overlooked as just basic household pets. Through this research, we've come to understand the deep level of thought that goes into those goofy dog behaviors we love so much.

However, what seemed missing in all this, what had gripped me and wouldn't let go, was the intuition that my family's bond with Kirby had reached deeper than just the natural and material. As a practical theologian, I have as my life's work the exploration of the spiritual significance of our lived, everyday experiences—and in my mind there are few experiences that pull you into the joy of everyday life more than caring for, and *being* cared for by, a dog. In the weeks following Kirby's death, I found myself wondering: Could there be something unique, maybe even intentional and holy, about dogs and their place in our lives? And if so, what soulful gifts could be ours to receive in the relationship, and what can we give? Could our connection with a dog in some way endure even into eternity?

Clearly, the very real grief *we* felt was evidence that we had bonded with this animal. Yet what about Kirby's perspective? Was he truly capable of loving us? Did he connect to us for reasons beyond the desire for food and shelter? And if this had been love between us, how deep did it go? Deep enough to touch the sacred?

I told myself it was nonsense—silliness, really—even as I ordered a book on the topic from Amazon. More likely than not, the curiosity I felt was nothing more than some kind of stage in the grief process, a process that would eventually resolve itself. At most, these questions might make for an interesting conversation with a good friend over a pint or two.

And perhaps I would have left it there—had I not discovered that one of the most famous scientists ever to study animals agreed with the premise of my search.

3

Bound by Spiritual Ties

I had to read the quote twice, it surprised me so much. Konrad Lorenz, a Nobel Prize–winning zoologist, was one of the first scientists to be intrigued by the unique qualities of dogs, refusing to overlook them as being too common to be worth studying. He examined all sorts of other animals, but for Lorenz, dogs were different: "The whole charm of the dog," he wrote, "lies in the depth of the friendship and the strength of the *spiritual ties* with which [the dog has] bound himself to man."[1]

When I read this, *spiritual ties* jumped off the page as if it were three-dimensional. I took a deep breath and put the book down for a second before lifting it and making sure I had read the phrase right. As a Christian theologian, when I think of human beings as spiritual, what I mean is that we are more than matter. We also embody an intrinsic yearning to connect to God, to seek transcendence—for a connection that pulls us beyond ourselves. We sense that there is a greater meaning or a deeper reality than what we can merely see or touch. Could Lorenz *really* have meant that this dynamic was present in our relationship with canines? This is what had been haunting me, but I hadn't been brave enough to admit it.

Much of Western theology and philosophy has made strong (and convincing, I believe) arguments for the uniqueness of human beings. To assume that a human being with extended consciousness could *spiritually* connect with a dog seems far-fetched against the backdrop of modern intellectual history. It wasn't hard for me to believe that a spiritual connection might work in one direction—that people project spiritual connections onto dogs. Maybe dogs are just blank slates that allow us to transfer onto them our emotions and longings. Yet Lorenz was saying something different. He was pointing to the possibility that something about dogs leads *them* to seek spiritual ties with *us,* that the very substance of our connection to our dogs is a two-way street.

Lorenz, an Austrian scientist in the second half of the twentieth century, pioneered the discipline of studying animals in their own habitat, believing that their full range of emotion and behavior was not able to be adequately observed in laboratory settings.[2] He was a kind of European Dr. Dolittle, whose experiments involved swimming with swans and having goslings follow him around on his home and farm as if he were their father.

Lorenz had deep appreciation for all animals and would not deny that elephants and dolphins have great intelligence, even possessing the ability to show empathy. Lorenz's research also led him to believe that dogs have a greater capacity for this kind of connection, that the human-canine relationship runs deeper than any other.

Even though it hardly *seems* like hard science, Lorenz's intuition made sense in light of my own dog's death. When Kirby died, I left the room. I was so surprised and overwhelmed by the grief and pain that I had to bow out. Yet my eight-year-old stuck it out. Maybe it's one of those mysteries of childhood, but my son, who sobbed for days after Kirby died, had this moment of total composure and absolute sureness of what needed doing in the face of death.

From the time of Owen's baptism, I have made the sign of the cross on his head and blessed him each night as I've tucked him in. We look into each other's eyes, and I trace the mark of grace upon him before he drifts off to sleep. This mark signals that he is freely welcomed and loved, unconditionally and without end, that nothing he can do, good or bad, can ever earn, or lose, God's love for him. It says, *You belong to God. Your life is held in something greater. You are known and claimed.*

This is the mark he bestowed upon Kirby in the vet's office, sealing our departed dog with a sign of God's never-ending love. Owen sensed something that eluded me. God, Kirby, and our little family were, in that moment, connected in a way that went deeper than I was willing to acknowledge. Owen knew he belonged to God; so, obviously, did Kirby. Even with, or perhaps partly because of, all my theological knowledge and study, I had become blind to something my boy was able to see and touch as if it were second nature.

Over those next days and weeks, I told myself that my nagging questions were an attempt to explain to Owen why he loved his dog so much and cried so deeply when he lost him. It felt ludicrous to assume that a dog who ate garbage and enjoyed drinking from the toilet could be spiritually tied to me. Still, Owen clearly had tasted that deep bond and knew what to call it. I wanted to understand it for myself. If a scientist such as Konrad Lorenz could see a spiritual tie between dogs and people, maybe a theologian like me could, too.

Coming Clean

One evening shortly after Kirby's death, I sat down with Kara and told her about my quest. I had this idea, this thing I wanted to explore. Maybe there was something there and maybe there wasn't, but just so she wouldn't be surprised when she got the credit card statement, I told her I was going to order a few more books on dogs. She looked at me sympathetically and patted me on the shoulder. I headed to the computer and ordered a stack of books from Amazon, following the trail from Lorenz to authors and scholars who'd built on his work in the years since.

Of course, none of the books said anything about spirituality and dogs—I hadn't expected them to. Instead, book after book presented scientific arguments exploring the evolutionary history, brain structure, and social patterns of dogs. The scientists represented in my collection of books chose to follow Lorenz's method of studying animals in their natural environments, but had little interest in venturing down the treacherous path of the spirit. I had a feeling I was moving into uncharted territory. Even Lorenz had made his claim without feeling any obligation to say more about it.

Sometimes a colleague asked me what I was working on, and I didn't know quite how to respond. It's normal, in conversation with fellow theologians, for one of us to say, "I'm working on the historical conceptions of reconciliation and how these have eschatological impact and therefore frame our political views of rights and privileges." For me to say in return, "I'm exploring our spiritual connection with, um, dogs," would be like a nuclear engineer saying, "I've decided to spend the next five years studying the inner workings of bottle rockets."

One day, I got brave and told a leading theologian about my research. "Spiritual connection?" he responded. "I have a golden retriever—he's nothing but a furry garbage disposal. Most days I figure if it were between me and a piece of grilled chicken, he'd take the chicken every time. Those big eyes and tail wags are nothing more than the long con that gets our dogs soft beds and fat bellies."

I nodded along, knowing where this was headed. I had doubts myself. Maybe I was overestimating the sophistication of dogs. Maybe cats were the true noble pets.

The theologian continued, gathering steam, "To be spiritual, you have to have a mind for something beyond you. You have to be able to transcend your own mind and seek the mind of another—God or a spiritual force or even another being. I really don't think dogs have such a mind. To believe a dog is spiritual is like trying to argue that cats are NASCAR fans."

He clapped me on the shoulder and walked away.

4

Mindless, Furry Machines

My dog skeptic theologian colleague had a point. I thought of those pictures of "dog shaming" that are popular on social media. You know the ones. In the photos, a dog looks at the camera with sheer guilt on his face, wearing a sign stating his offense: "I Ate Two Sticks of Butter and Smeared the Third on the Couch." Or, "On Walks, I Roll in Dead Stuff and Then Jump into Mommy's Bed When We Get Home." Or, one of my favorites: "I Sneak into the House of Our Buddhist Neighbor and Eat Their Food Offering to Buddha."

Of course, what makes these pictures so funny is that we all know dogs who routinely commit such weird, gross, destructive crimes against their owners. There they are: caught, guilty as charged, incorrigible, actually, and with all their dopey lovability intact. This completely reinforces the professor's point: dogs are con artists.

Had Kirby merely duped us into loving and feeding him? If he had, my own grief would be a little bit pitiable and maybe even silly, on the order of burying a stuffed animal or retiring an old, beloved car to the backyard. But if I was going to go anywhere with this project, I had to address the assumption that dogs operate on instinct alone. To know what might be happening from the dog's side of the relationship, I had to start with the mind of the dog.

The Rational Soul

Knowingly or not, my theologian friend was following the thinking of the seventeenth-century French philosopher René Descartes, chief philosophical architect of the Enlightenment. As a young boy, Descartes watched Europe burn as Catholics and Protestants fought and killed one another during the Thirty Years' War. A particularly clever young man, he sought to make war obsolete by finding some common ground of religious or moral ideology that all people could agree on.

Descartes's studies led him to conclude that nearly everything a person might know or believe, including our own consciousness, could be doubted. You might believe you're reading this book right now, holding it in your hands (maybe) as you sit in a comfortable chair.

On the other hand, you could be just dreaming, sliding into a hallucinatory state. The only certainty, according to Descartes, is our own existence. When you doubt the reality of your book or your chair, you can at least be sure that *you* are the one doing the doubting. Doubting itself is proof of your existence. This led to Descartes's famous dictum "I think, therefore I am." You *are* because you have a mind that enters into such deep, rational contemplation that you are able to doubt everything.

Descartes believed he had done more than discover a solid foundation on which both Catholics and Protestants could agree—more than that, he had discovered the shape of the soul itself. The soul, to Descartes, is this conscious ability to reason, and it is uniquely human. Because it seemed clear that animals were unable to do this doubting and reasoning, Descartes assumed that they are soulless, able to perceive and sense, but essentially machines made of muscle and fur.

When I think of Descartes's view of animals, I remember the classic *Seinfeld* episode in which Kramer starts taking the same medication as a dog he meets in the park. Kramer distrusts doctors in general, but because he and the dog have the same cough, Kramer borrows the dog and takes him to the vet in order to obtain the meds the veterinarian prescribes for the dog. As the episode continues, Kramer begins to act more and more like a dog, exhibiting increasingly absurd, hilarious behavior.

In 1989, the first year *Seinfeld* aired, Michael Richards, the actor who played Kramer, played a character named Stanley Spadowski in the movie *UHF*. Spadowski is slow and dumb, getting laughs through his inability to keep up. Later on, when Richards was cast as Cosmo Kramer, he initially played him just as he had Spadowski: slow, dumb, and easily duped. By the second season, though, Richards says he realized that Kramer was nothing like the mindless Spadowski. Spadowski is a step (or six) behind everyone else, and that's what made him funny and lovable. Kramer, though lacking in self-awareness, is actually a step *ahead* of everyone else. He is as brilliant and resourceful as he is odd, and in the end he always proves loyal to his friends. Kramer comes across as dumb because he is so unconventionally mind-*ful*. That's what makes him such an iconic character.

It could be that dogs are like Spadowski, dull creatures programmed to beg and wag a tail to survive. Or they might be more like Kramer, unconventionally and peculiarly mindful, reacting to us at times before we are even aware of our own actions.

I was stuck with an either/or dilemma. If forced to choose between dogs being Spadowskis or Kramers, I would have to say that the history of philosophy and the assumptions of religious thought would be on the side of the mindless Spadowski. Descartes believed that without a mind that could reason, there was no soul. Animals would then be less like human beings and more like biological apparatuses. Descartes believed this so fully that he went so far as to argue that the howl of a dog is no different from the screech of the brakes on a carriage.

By this line of reasoning, dogs may *seem* to respond to us, but this is only because they have been programmed to do so by their innate need to survive. To say you know your dog loves you because he's happy to see you when you return home is as logical as saying your iPhone loves you because it knows what time to wake you up. Your phone has no thoughts or feelings for you; it simply performs the functions it has been programmed to do.

Could it be that our beloved dogs are no different?

All We Are Is Programming

Descartes's perspective that animals could be thought of as mindless, programmable machines was bolstered by the findings of physiologist Ivan Pavlov. In 1901, Pavlov famously discovered that, with the right conditioning and stimuli, you could, quite literally, program a dog.

Pavlov got the idea for his best-known experiment when he noticed that his dogs drooled in anticipation whenever he brought them food. This seemed to be a preprogrammed response, in which the glands that produced saliva turned on like faucets when the dogs smelled and anticipated their dinner. When Pavlov began studying this phenomenon, however, he was surprised to learn that he could actually reprogram the dogs' behavior by ringing a bell before the food arrived. He'd ring the bell, then start getting them their food, slowly increasing the time between the sound of the bell and the arrival of the food. Eventually, the dogs could be programmed to launch into eating mode just by hearing the sound of the bell. So, not only were dogs instinctive machines, Pavlov discovered, but their programmed instincts were malleable. With the right behavioral conditioning, their behavior could be rewritten.

In other words, Pavlov's legacy was to prove that there was no mindfulness involved in dogs' behavior. Behavior is produced by connecting positive stimuli or negative stimuli with acts you want to encourage or discourage. Almost all dog training works from this theory. Want a dog to learn to sit? Give it positive stimuli when it sits, and you will program it to sit when told. Want your dog to stop pulling on the leash during a walk? Produce negative stimuli (corrections), and you will reprogram it to stay near your heel.[1]

Pavlov seemed to have discovered that your dog isn't a step ahead of you, like Kramer. More likely, your dog is a step behind you, like Spadowski—slow but trainable.

Picking up on Pavlov's finding, the psychologist John B. Watson offered a perspective on animal behavior called behaviorism. Through Watson's charismatic disciple B. F. Skinner, behaviorism became the primary way of thinking about animal conduct—and even human psychology.

B. F. Skinner was an American psychologist whose

beliefs were shaped by intense experiences in his child-hood. Skinner's religious grandmother often spoke of the fires of hell. When Skinner was eighteen years old, his sixteen-year-old brother died of a cerebral hemorrhage, and Skinner became an atheist when a Christian teacher tried to mitigate his fears that his brother had been sent to hell. Having come from this kind of background, it's not difficult to see why behaviorism would have appealed to Skinner.

For Skinner, all animals and humans *are* is behavior. Any belief that a higher consciousness has to be at work is simply an illusion. Behaviors build on one another; each behavior connects with another behavior, LEGO style, to build a response system. It is the stimuli of reward and punishment that lead us to change, or to connect our behavior with the good of another.[2]

A dog, then, in Skinner's view, is not much more than a furry TiVo with an appetite. Just as your TiVo connects your watching of *Pawn Stars* with a show you've never watched (e.g., *American Pickers*) because it has connected one behavior with another, so Pavlov's dogs connect the stimuli of the bell with eating, and they behave by drooling. The more you use your TiVo, the more you program it with stimuli data and the more it will behave as you want, giving you the illusion that it actually has a mind. According to Skinner, our dogs work in the same way. When your dog is happy to see you at the end of the day, it's not because he really cares about you. It's because he's hungry.

A Dead End?

Let's just say that my research up to this point was not granting me glimpses of heaven in a hound. Descartes, Skinner, Pavlov, and the dreary dog skeptic theologian all agreed: dogs are food-motivated beasts who are shaped by base stimuli and responses.

Had I arrived at the end of the road for high-minded dog lovers? Was I nothing more than a deluded romantic, trying to read more into Kirby's behavior than could reasonably be found there? If so, I would have to conclude that Kirby had no mind for Owen or Maisy or Kara or me. There was no spiritual two-way street between us and our pet dogs. All that we had experienced with Kirby—the companionship and play, the years-long bonding with our children, the sacredness of his send-off—were simply projections of our human longings.

Still, I wasn't ready to concede. The stream of thought that the theologian pointed me to, the river that ran from Descartes to Pavlov to behaviorism, just didn't satisfy. It gave me little insight into why it hurt so badly to lose Kirby, or why it seemed our family had entered into a shared reality not only with our dog, but with others who had lost dogs. It struck me as ultimately reductive.

I suppose great minds like Descartes or Skinner would have reminded me at this point that just because you *feel* something doesn't mean it's true. Descartes would actually have encouraged me to doubt my feelings and turn to more rational explanations. Yet I couldn't.

I couldn't forget, for example, the memory of one Christmas we shared with Kirby, and the simple yet profound theology expressed then by another great mind: Owen.

5

It Happens Only Face-to-Face

The day after Kirby died, Owen taped a picture to the headboard of his bed. It was a photo of him as a baby, dressed as Santa Claus, held up to look as if he were riding Kirby, who was looking sheepish in a pair of foam antlers. The picture was our Christmas card the year Owen was born. On the edge of the photo, Owen had written, "My best friend." As I looked at this photo, the ideas of Skinner, Pavlov, and Descartes began to feel too one-dimensional to account for all that richness.

There was more. Every night after Kirby's death, Owen pleaded with me to pray that God might give him a dream of his dog, to take him one more time into an encounter with his best friend. It wasn't just the lack of a dog that Owen was feeling. This had to go deeper. I was sure of it.

In the preceding chapter, we asked a simple question:

Does the behaviorist model (stimulus begets response) explain what Owen and dog lovers everywhere interpret as affection and relationship? Remember, we're on a quest to identify the spiritual or higher capacity of the dogs we love, if that capacity exists.

Our answer so far is a qualified yes. But, it's also no, I don't think so. Yes, our dogs are predictably motivated by the food and attention we provide. But also no, because there's more going on. A whole, other world is going on, in fact, a world of feelings, memories, behaviors, and meanings for both dog and owner.

In this chapter, I want to show you research that has helped scientists see past the furry canine machine model and find in dogs' responses to humans a fascinating affinity, and a visceral and unfailing preference, that's not to be found in any other animal.

A Smackdown for Behaviorism

Over the last half century in the scientific community, Skinner and behaviorism have taken it in the teeth. In 1959, against the backdrop of Colonial-era houses and winding cobblestone streets in Princeton, linguist and philosopher Noam Chomsky was asked to write a review of Skinner's 1957 book, *Verbal Behavior*. Chomsky's crippling assessment argued that certain core realities, such as language in human beings, could not be explained by behavioral stimuli. The idea of creatures as mindless operating systems, he stated, simply couldn't explain human realities such as creativity, language, empathy, bonding, and play. This review launched Chomsky's now-legendary career in linguistic studies, and exposed behaviorism as being inadequate for explaining human beings. Language alone reveals that we are far too mindful.

More adequate theories on animal behavior have been proposed.

Chomsky's ideas were limited to human beings, but in animal studies, a perspective born from odd old Konrad Lorenz called cognitive ethology asserted that, in opposition to behaviorism, an ability for complex thought does exist in animals such as elephants, dolphins, and apes. Popular dog writer John Homans says nicely, "Their minds . . . solve problems, map landscapes, [and] distinguish friends from enemies."[1] For example, we now know that dolphins have a sense of the individual self as separate from others. Elephants grieve the death of a member of the herd. Animals may have minds very different from ours—no spider monkey has ever written a book on his species' spiritual connection to bananas—but research in cognitive ethology has discovered that animals nevertheless *do* have minds and cannot be understood as merely dim-witted, stimuli-response mechanisms.

The "cognitive" commitment in cognitive ethology contends that understanding animals' minds may help us not only appreciate them more but also better comprehend the origins and structure of our own minds.

The "ethology" commitment in cognitive ethology asserts that because they have minds and are more than a collection of instincts, animals must be observed in their own natural surroundings.

This leads to this important insight: When it comes to our dogs, this natural environment is, uniquely, the human household. Yes, there are millions of dogs in the world whose natural environment is not the human household; feral dogs roam the streets all over the world, and strictly working dogs know only kennels. Yet the natural habitat of dogs who have friendships with eight-year-olds, the dogs I wanted to know about, is the human home.

If dogs are indeed more than a bundle of learned responses, then understanding their deeper capacities will happen only through observing them in their natural surroundings: our backyards, our couches, and the backseats of our cars.

When Oreo Went to Emory

In just the last decade and a half, breathtaking discoveries have been made by cognitive ethologists studying the dogs who live with us. Take Brian Hare of Duke University. He cut his scientific teeth in the late 1990s studying under a renowned scholar named Michael Tomasello, his professor at Emory University. Hare was captivated by Tomasello's search for the evolutionary link to human language. He believed that if this link could be spotted, we would know more about the evolution not only of our own minds, but animals' minds as well.

At the time Hare met Tomasello, the professor was running experiments with great apes, to see how they responded to human gestures, seeking to determine whether language might have sprung from the ability to read one another's movements as communication. Tomasello was discovering that apes (our closest living evolutionary relatives) did quite poorly at recognizing human gestures.[2] Researchers could use behavioral techniques to teach apes to understand gestures such as pointing, but left to their own devices, the apes seemed uninterested in reading one another's communicative movements, let alone the gestures of humans.

Drawing on this research, Tomasello concluded that the ability to read gestures was a uniquely human capacity, the capacity that had produced our distinctive penchant for language. Hare, Tomasello's high-spirited undergraduate assistant, disagreed.

Hare pointed to his family dog, Oreo. He thought Oreo was able to read gestures quite well. Tomasello, who had millions of dollars in grant money and pens full of chimps and great apes, thought this absurd. Why would an average house dog be able to solve a mental problem that his vaunted primates couldn't? Still, Hare was convinced that Oreo could.[3]

You see, when Hare was young, he spent hours in his big backyard throwing baseballs, with Oreo serving as his partner and retriever. Hare told Tomasello that he could signal his dog to start running with a motion of his finger. A mere twitch of the foot or elbow, even just a glance in the direction of the ball, was enough to launch Oreo across the yard.

Tomasello remained skeptical. So Hare began constructing an experiment that he would continue to adapt throughout his career, to prove that dogs indeed understand gestures. He placed two cups about two meters apart, pretending to put a treat under one and actually putting a treat under the other. Then he would stand back and point to the cup hiding the food. Each time Hare did this, Oreo—and eventually the many other dogs Hare would come to test—followed the pointed finger. Hare then blindfolded the dogs to keep them from watching as he put the treat under a cup, giving them no visual cue as to which cup hid the food. He would then take the blindfold off the dog and point. Nearly every time, the dog gave its attention fully to the face of the researcher, awaiting a signal, ready like a sponge to soak up the researcher's gesture—and then responded to the pointing.

As Hare continued ramping up his experiment, he had to control for smell, lucky guessing, his own body language, and other possible explanations for the dog's success. Yet at nearly perfect rates, all the dogs studied the point of the researcher, and did so right from the beginning. There was no sense, as there was with the chimpanzees, that following gestures was a learned behavior.

Hare explains: "Maybe during many interactions with me, Oreo had learned to inflexibly use a few signals. If this was true, he should show improvement during the tests and would have trouble reading cues he had not seen before. But Oreo [and the other dogs] almost always chose correctly on the first trial and did not show improvement . . . because he was nearly perfect from the start."[4]

Hare and Tomasello then experimented with chimpanzees who had learned to read a gesture like a finger point. Still, even though they had been trained to understand pointing, the primates, unlike dogs, seemed to have no ability to understand new impromptu gestures. Even more, Hare found that puppies did as well at his experiment as grown dogs. The reason began to come clear. The scientists realized that they weren't testing the *learned behavior* of dogs, or their readiness to read physical gestures, but something else: an *innate attraction*.

Hare had discovered something big: Dogs have an inborn mind for human connection. From a very young age, they express this attraction by where they look.

And where they look is to the human face.

The Difference a Gaze Makes

Like no other animal, dogs seemed to be innately wired to look to the face of human beings for direction and connection. Dogs were the only animals tested who responded like human infants, seeking cooperation by attending to the human face. In fact, Hare even discovered that if the person pointed his or her head in the direction of the correct cup, while not looking directly at it (that is, while gazing beyond it, giving it no direct attention), the dog would read his or her intention and respond instead to the finger pointed at the other cup.[5] With almost spooky sophistication, the dog could tell where the researcher was putting their attention and what part of their body was communicating. Reading that the researcher's face was looking beyond the cup, and therefore providing no information, the dog would take clues from the pointed finger.

To measure whether perhaps gesture reading came from dogs' pack animal nature, Hare also tested dogs' closest relatives, wolves. The wolves, however, failed the test.[6] Instead of acting like domestic dogs, they acted just like chimps, giving little to no attention to human beings, ignoring the human gestures and having no concern for the human face.[7]

Other cognitive ethologists were corroborating Hare's results. A group in Hungary did an interesting study in which they put a piece of raw meat inside a metal box. Making sure the wolf or dog knew that the meat was in the box and could smell it, they allowed the animal to try to retrieve and eat it. The trick of the study was that the animal was not able to get to the meat on its own.

When the wolf performed the experiment, it would walk over to the box, sniff, and then dig, seeking to free the meat. When digging didn't work, the wolf was smart enough to look for other options to free the savory snack on its own, such as knocking over the box or getting at it from other angles. Yet, the wolf never did what dogs do naturally. When the dog performed the experiment, it first went to the box and dug hard for the snack—but when the snack wasn't released after a few seconds, the dog sat next to the box and turned its attention to the researcher. The dog gazed directly at the researcher, as though asking for help.

As you might guess, a wolf has a larger brain than a dog, and is smarter at finding its own solutions to problems. Yet the wolf has no mind for us.[8] Psychologist and dog theorist Alexandra Horowitz explains the difference this way: "Dogs look at our eyes. If this behavior is unsurprising, it is because it is so human: we look. Dogs look, too. Though they have inherited some aversion to staring too long at eyes, dogs seem to be predisposed to inspect our faces for information, for reassurance, for guidance."[9]

At our house, we know what Horowitz is talking about. (Maybe you do, too.) Kirby had what Kara always called his "question mark" eyes. He'd sit still as a statue, staring at, say, my face as hard as he could, eyes as wide as saucers. Once he had my attention, he'd take his eyes from my face and move them to a tennis ball, and then return them again to my face, and then back to the ball. Kirby was soundlessly screaming at me to throw the ball, to come on and play with him, to be with him.

Like infants, dogs are able to read the gestures of people, and to gesture back, forming a bond of cooperative communication. Dogs will stand at the door if they need to go out, or sit in front of you when they're ready for dinner. When it seems the right time of day for a walk, a dog might even bring you his leash or pounce on you and lick your face to compel you to recognize that her routine is calling.[10]

It's not uncommon to see a dog owner stop in the middle of scolding his pet and physically turn the dog's head back to make eye contact. Why? "We want dogs to look at us when we are talking to them," Horowitz writes, "just as we use gaze in human conversation, in which listeners look at the face of the speaker more than the reverse. [That's why] we call their names before speaking to them."[11]

Research on human connection bears this out. As I write this, there's a video going viral on the Internet that I'm sure, soon enough (like tomorrow), will be replaced by another Chewbacca-masked mother or something of that sort. But for these twenty-four hours, it's making people cry all over their keyboards. It's shot in an empty warehouse. Pairs of strangers, one of them a refugee, enter the room. They are directed to sit across from each other and, in silence, to look at each other's faces for four minutes. The video is a social experiment based on the psychological research of Arthur Aron, who argues that "four minutes of eye contact brings people closer to each other than everything else." And the video proves it. Face-to-face encounter within minutes brings tears, laughter, joy, and shared pain. These four minutes see children become friends, and two people find attraction and loving connection. The power for face-to-face connection brings an encounter of spirit.

This is why it's not surprising when you hear, as I have, a hardened outdoorsman confessing something like "I've cried only twice in the last thirty years: when my dad was diagnosed with terminal cancer and when my hunting dog needed to be put down." The hunter mourns the death of his dog because a deep connection is now lost—a visceral relationship that began, like a close human relationship, in long and careful attention to the face of the other.

I Only Have Eyes for You

Before we move on in our canine soul search, I want to make an important distinction: the kind of attentiveness, or depth of communication, we've been talking about in dogs is not primarily dog to dog, or dog to any other species, but dog to human.

This fact is truly remarkable.

Dogs, like many other mammals, become attached to one another, forming packs. Dogs are even known to make friends with cats, turtles, and birds. Yet what their hearts long for most is to connect with people.[12] Scientist Vilmos Csányi states, "With well-designed experiments we can even show that puppies are attracted more powerfully to humans than to their own species."[13]

In one study done at a shelter, it was discovered that, on average, it took several days for a dog to attach to another dog. Yet, in comparison, it took just hours for a shelter dog to connect to a human being, and this connection with a human being was almost always stronger, and preferred by the dog. Nearly every time, given an option, the dog chose to be with a human over a fellow dog. "It's a trivial and obvious point," noted Tomasello after seeing Hare's study, "but dogs aren't doing this with other dogs. They're doing this with us."[14]

I couldn't help but see deep theological possibilities in this fact. Dogs are unique because of their ability to be drawn to our faces, and Christianity and Judaism are faiths in which the human face holds great importance. For example, one of the central benedictions in Protestant churches is taken from Numbers 6:25, "The Lord make His FACE shine on you, and be gracious to you." And the Psalms are filled with verses such as "Do not hide your FACE from me in the day of my distress" (Psalm 102). In 1 Peter 3:12, in the New Testament, we read, "For the eyes of the Lord are toward the righteous, and His ears attend to their prayer, but the FACE of the Lord is against those who do evil."

All throughout the Christian scriptures, we read that God has fully revealed Godself in the *face* of a first-century Jew, and that what God wants from us is that we turn our face toward him and toward one another. To live the Christian life is to be always attentive to others, seeing the face of Jesus in the face of your neighbor, as

Jesus says in Matthew 25. And here was science, saying that dogs, these beautiful butt-sniffers, are uniquely shaped to do for us exactly what we're called to do for one another.

I began to wonder if Owen might be more right than the theologian.

6

Window to the Soul

I teach at a Lutheran seminary. The stairwell up to my office has a spooky death mask of Martin Luther mounted on the wall in a glass case. You should know: my theological tradition isn't known for touchy-feely ideas such as attributing spirituality to animals. We're not Saint Francis folks; we're Martin Luther and John Calvin folks. The Reformation that Luther and Calvin led was a rebellion, for sure, but one that emphasized order, personal responsibility, and discipline. We're not called the Frozen Chosen for nothing.

However, it's also true that Luther himself was a known dog lover. Once, he told of receiving a letter from a friend describing how a dog "pooped in the grave of the Bishop of Halle," which prompted Luther to tell his own story of a formal processional with banners around a Catholic church. When the verger set the pot of holy water down on the ground, a local dog approached the processional, lifted his leg, and began peeing into the pot. The priest, who had been in the act of sprinkling the water, bellowed, "You impious dog! Have you become a Lutheran too?!?"[1]

If Luther had had a Facebook profile, along with posts about the hypocrisy of popes and the delights of drinking beer, his Timeline would have revolved around his dog, Tölpel. He loved spending time with Tölpel, expressing to his friends that he wished he could pray with the same concentration with which the dog begged his master for meat. Luther, whose theology saw God *not* in the pomp and circumstance of the religious, but in the humble and lowly, said, "The dog is a very faithful animal and is held in high esteem . . . Our Lord God has made *the best gifts* the most common."

We know that dogs seek to communicate with us, but is it insane, is Luther drunk, to think there is a spiritual dimension or potential to this? Can something so common as a household dog draw us into the spiritual? Could it be that in these common four-legged creatures, God has given us a beautiful gift that's meant to encourage our highest forms of spiritual longing?

We usually think of spirituality or the soul as being defined by the ability to make rational choices, to believe in something beyond yourself (or anything at all); this is the legacy of Luther and Calvin's Reformation. But what does that mean for dogs, when they can't hold theological beliefs, choose to be Lutheran, or make a conscious ironic theological statement for the priest or verger (or deceased Bishop of Halle) to interpret?

Hare's studies prove that dogs are able to attune to us with amazing skill, reading us and desiring to be with us. It was interesting science—and I could tell Owen that Kirby was amazing at reading him—but so what? This doesn't necessarily prove that they were friends. It doesn't explain why losing him devastated me so much. It doesn't name the inexplicable quality beyond the natural or material that seemed to be present in our relationship with Kirby. And I'm not about to assert that a Goldendoodle could believe in substitutionary atonement or God's providence.

So, what does gesture reading and our dogs' long looks into our faces have to do with spirituality? How do they echo the core theological commitment that God comes to us in and through relationships that are face-to-face? A lot, actually.[2]

The Core of the Spiritual

Spirituality, or spirit, or even *soul,* is often assumed to be some kind of ethereal, invisible thing inside us. We assume it is like the Force in *Star Wars,* some supernatural power that gives us wisdom like Yoda's and magic like Obi-Wan's. I suppose in some ways this is true, but spirituality can also be seen as an awareness of something that transcends you—a relationship with something that is beyond you. Particularly in the Christian and Jewish traditions, a spiritual connection is a deep sense of sharing in the life of another person, whether it's God, your spouse, or a close friend. This sharing of life is so deep and beautiful that we often call it love.

The Jewish thinker Emmanuel Levinas claims that humans are infinite spirit—but not through a disembodied substance hidden in our chests, as Descartes believed when he took his poor wife's little dog and cut it open looking for the gland that contained the soul. (Spoiler alert: he never found it!) Rather, Levinas asserts that the spirit or soul is experienced through our faces, which invite us to share in one another's being. The human infant can read gestures early on because the infant is uniquely drawn to the face of her caregiver, and the caregiver cannot take her eyes off the face of the baby.[3] It is seeing and responding to this deep need for connection, offered through the face of another, that makes us spiritual.[4]

Levinas noticed, while in a prisoner-of-war camp, that encountering the face of another seemed to place a particular call on a human. It led us beyond ourselves, to suffer with others, to protect them. To cage and destroy human beings was a way to say, "Your face (being) doesn't matter, and I don't need to see it."

A friend recently shared with me the story of his grandfather, who is suffering from severe dementia. In the midst of this crippling disease, his wife of forty years died. At the time of her death, he could not remember her, but after three weeks of her being gone, he said to my friend, "I am missing a face that I can't remember."

One reason I believe humans are spiritual beings is because, as a species and as individuals, we are compelled to reach for transcendence—to seek (I believe) an experience of God in art, in worship, in nature, and in the company of our neighbor. In the Christian traditions, to be "made in God's image" is to be made for the purpose of knowing others and being known. It is this innate draw that causes us to read the faces of others, from our first days, reminding us that we are more than animal, vegetable, or mineral. For me, the best word to describe this capacity is *spirituality*.[5]

If face-to-face encounter is a doorway into the spiritual, it makes good sense that we feel a spiritual connection to our dogs. To have a dog joyfully pant at the window when your car pulls into the driveway, or follow you around as you make your morning coffee, or look up into your eyes before laying her head on your lap, is to feel seen and connected to another. Yet what, exactly, is the purpose of this connection?

What's the Motive?

To read others and connect deeply is a beautiful and powerful thing, but this capacity doesn't mean that every relationship we have is spiritual, or even productive. Sometimes people connect with us to hurt us, to deceive us, to use our spiritual capacities in some way against us. While my friend the theology professor was wrong about dogs being mindless, perhaps he was right that they are just con artists. Are dogs like the handsome, gold-digging hunk who woos the older woman with his sensitivity and openness only to gain access to her bank account?

It's very clear that dogs are able to attune to our personal frequency, but if this is truly spiritual, then it has to be deeper than just tuning into us. What is their motive? If they've merely found a way to connect to us for their own reasons, for survival or for gain, then it doesn't seem mutual, transcendent, or spiritual. Do they really want to be with us for the sake of being with us? Do they enjoy our presence and seek experiences that don't necessarily benefit them functionally? In other words, are dogs smart enough simply to exploit our spiritual proclivities, or are dogs actually *part of* our spiritual connection with God and others? Is this experience somehow mutual, a two way street? Dogs seek our faces and read our gestures, and this pulls deeply at the spiritual cords of *our* being. We feel drawn toward the experience of transcendence, but does the dog?

It was to these questions that my search now turned.

7

The Surprising Power of
Canine Compassion

It happened right after one of those announcements that travelers dread: "Hello. For those here at the gate waiting for Delta Flight 1974 to Minneapolis, I'm sorry to say that we're having an equipment issue, and your aircraft is still in Salt Lake City. Looks like a three-hour delay at least. We apologize for the inconvenience, especially right before the holidays, but we'll wheel out our complimentary snack cart and get you something to eat while you wait, okay?"

Groans went up all around me. The frustration among the travelers was palpable. We traded angry opinions about the airlines and their miserable complimentary snacks. We bemoaned our bad luck and ruined plans. The world that day seemed a cruel and uncaring place—that is, until a brown-and-white spaniel with floppy ears and a big doggy grin trotted into view.

The spaniel was wearing a "Therapy Dog" vest, and we all watched as he accompanied his owner with bold assurance up to the gate agent. With every step, the dog seemed to be telegraphing an important message to everyone at Gate A32: *It's a good day, folks! Everything's gonna be just fine.*

I watched as, like reverse contagion, anger and frustration began to melt from people's faces. Two children approached the dog and asked to say hi. The man consented, allowing the dog to shake their hands and show off some tricks. Within ten minutes, the dog was a celebrity in our corner of the airport. A small crowd huddled around the spaniel and its owner, talking and laughing, swapping stories. Something had happened among us. In a mysterious way, that little dog had defused a tense situation and replaced frustration with a spirit of kindness.

I wondered if Delta noticed. Sure, the free snacks had been a good idea, but what they really need for times like these is a Delta Dog. The Delta Dog could come to your gate, bestowing unconditional welcome and love; allowing calm, laughter, and playfulness to return to the crowd; reminding you with his doggy goodness that even though you are delayed, it's still very good to be alive, you are still human and still, in his eyes, delightful! (Surely Delta will want to give me several hundred thousand bonus miles for this brilliant idea.)

I was already knee-deep in exploring Lorenz's idea when that spaniel became a celebrity at Gate A32. As I watched him and pondered his impact on us, I was carrying in my backpack the eminent social theorist Robert Bellah's most recent work, a brick of a book called *Religion in Human Evolution,* which he'd published just before he died. The book, about the evolutionary origins of religion, had nothing to do with dogs, but it did explore the experiences that Bellah believed to be foundational for spiritual encounters.

In the previous chapters, I propose what "spiritual" *is.* I describe it as a deeply relational awareness or encounter, one that can take us outside our material preoccupations and into a profound awareness of others and of God. We experience this spiritual reality most often in face-to-face encounters. There is *something* about us or within us that is moved by being seen, but what is this *something*? What is the character of these face-to-face encounters, and more specifically, why do we assume them to be spiritual?

I needed to find out. It is one thing to show how dogs attend to our faces, but it would be quite another if this attention triggered in us (and them!) emotions that moved toward love.

This is where that brick of a book in my backpack came in. In it, Bellah proposed that through our capacities for empathy, bonding, and play, humans come uniquely hardwired for the religious impulse. Bellah was saying, in a mere six hundred pages, that:

- **Empathy** is the deep experience of feeling the other, such as tearing up when someone else cries or smiling when another person laughs. You actually feel, at some level, what he or she is feeling.

- This leads to **bonding,** in which we feel our lives actually tethered together. Our most basic experience of bonding concerns a parent to a child, and shared experiences such as rituals, trauma, and other circumstances that write our lives together bond us to others.

- **Play** is the energy that moves us near to one another. When we take a break from the demands of everyday life, we enter a space in which we can see each other anew. For a mother with her child, playing is both the way into, and the witness of, a shared bond.

Standing there at the gate, I was struck by how dogs, too, seem unusually wired for all three capacities: empathy, bonding, and play. And if dogs are, not only would the corollary be amazing, it would more clearly justify our calling the relationship we have with our dogs "spiritual."

I decided to look first at empathy.

What Does the Fox Say?

In 2013 a Norwegian comedy duo gave the world a song so ridiculous and therefore so catchy that it became an unexpected juggernaut on the *Billboard* charts, making what was intended to be a comedy song into a mainstream hit. "What Does the Fox Say?" was a song about the sounds animals make that culminated in a chorus based on a response to the title's question. Naturally, the comedians were interviewed incessantly about the song. They explained that they wrote it because it was the stupidest idea they could think of, making it, of course, very funny. But with tongue firmly in cheek, they continued to assert that the song asks a deeply mysterious question: what, indeed, does a fox say?

This silly song became the anthem of our house not long after Kirby's death. In a bizarre way, it served as a doorway out of the heaviness of our loss—and Maisy was patient zero. She'd heard the song in her first-grade gym class and sang it at what seemed like every waking moment, quickly spreading the contagion throughout the family. She'd shout, "What does the fox says? Ring-ding-ding-ding-dingeringeding! / Gering-ding-ding-ding-dingeringeding!" (If you somehow managed to avoid hearing the song, and I've just caused you to look it up online, I'm sorry.)

Silliness aside, it turns out that in the last few decades the fox has had much to say about the dog and the role of empathy in human-animal relationships. In the late 1950s, a Russian fur farmer began an experiment that has fascinated scientists for years. Dmitri Belyaev decided to begin breeding foxes, hoping to benefit financially from the great demand in Russia for their beautiful coats. Yet, Belyaev was also a gifted biologist, so while breeding these foxes for profit, he also began a simple test, one that revealed something surprising about dogs.

Before diving into that study, however, let's take a quick look back at the history of the domestication of animals, and dogs in particular. Even though genetics tells us that dogs share 99.6 percent of their DNA with wolves, I can tell you it's hard for the Root family to look at Fluffy, our neighbor's Bichon poodle, and believe she shares even a smidgeon of DNA with a gray wolf on the Canadian tundra. Yet that very disconnect actually shows just how quickly and recently, relative to the evolution of most species, dogs have come to look and behave so differently from wolves.

Nearly all dog theorists and scientists believe that this change in the dog's appearance and psyche is so radical (and late) that great alterations must have happened *during the process of domestication.* I'll try to explain.

Domestication is typically understood as the process of an animal or plant being adapted for human use. Pigs or corn, for example, have been bred to serve our purposes. We have forced them to evolve in ways that make them useful to us, not to their own well-being as a species.

Some Jewish and Christian interpretations of Genesis 1:26–28, in which humans are blessed to "have dominion over" all living things, are used to support this form of domestication. Humans are mandated, according to this view, not to be in relationship with creation but to control it. Yet many theologians have argued that this interpretation of dominion is a misreading, and that humans, as bearers of the image of God, are to be stewards of creation, to befriend it and care for it as a mother does her child.[1] This kind of befriending relationship describes more closely the way dogs became domesticated. Unlike pigs, who seem to have been domesticated outside their own will, most dog theorists believe that the wolves that became dogs *domesticated themselves.* They *chose* to live with human beings.

Here's where this gets really interesting. It seems that for good or ill, when an animal gets close to human beings, it changes in mind and body. Often this is simply due to the will of humans and our desire to use animals for our own ends. For example, once domesticated, pigs and cows were unafraid of human contact and became larger than their wild counterparts, providing us more meat to eat.

The wolves that became dogs also greatly changed in connection to us. They now come in all shapes and sizes, like the proud pug, the water-loving Chesapeake, the lap-friendly teacup Pomeranian, the regal Great Dane, the sonorous beagle, and hundreds of other breeds. What's interesting is that most dog theorists believed that, unlike pigs and cows, the initiative for domestication came not from us, but from wolves themselves. *They* approached *us,* hanging around our camps. We did some selecting, but not for functional reasons; for relational ones. We chased away the mean ones and took the kind ones further into our lives, and these kind wolves *quickly* became obsessed with us, wanting nothing more than to be with us.

This was all just speculation until the fox had its say.

The Friendly Fox

This takes us back to the Russian fox breeder Dmitri Belyaev. Foxes are notoriously frightened of humans, wanting almost nothing to do with us. Yet Belyaev noticed that a handful of the foxes he was breeding for their fur were actually less frightened than the others. So he separated these foxes, allowing them to breed with each other only, doing experientially with foxes what we believe happened at the beginning of domestication with dogs. Belyaev continued breeding for only one trait—not fur color, skill, or usefulness, but tameness toward human beings. The foxes that were kind and gentle to people continued to be bred with one another. Within just twenty generations, huge transformations occurred in both the bodies and minds of these foxes.

The foxes that were kind to people mysteriously lightened to a grayish silver, losing their black/silver coats. After a few more generations, their ears became floppy and their tails curled. It wasn't the color or the floppy ears that led Belyaev to breed one fox with another. It continued to be only the one trait of tameness, but with every generation, this breeding pattern transformed their appearance. *Their relationships with human beings were changing their very shape.* Yet, if this wasn't shocking enough, it was the transformation of the mind that further intrigued scientists.

As the generations of foxes were changing color and their ears were dropping, not only did their tentativeness fade, but these foxes became infatuated with people, acting more like puppies than wild animals. If a person walked near a normal fox, the fox would run in the other direction and hide in the corner of its cage to escape the gaze of a human gawker. The foxes bred for tameness, however, soon began not only to approach people, but to be enamored of them, exuberantly seeking to be near them and touched by them and wagging their curly tails at the sight or sound of people. This was very unfoxlike behavior. In just a few short generations of domestication, the fear of people was quickly replaced with an almost addictive yearning to be near them. These silver foxes lived for human touch.[2]

Belyaev had succeeded in domesticating a population of wild animals simply by breeding for friendliness toward people; all the other traits appeared as by-products. Soon some of the silver foxes even began to bark, leading scientists to believe that they had discovered the reason that dogs bark. Just as normal foxes don't bark, neither do wolves.[3] It appears that barking is an (annoying) trait of domestication, but one that serves the purpose of getting people's attention, a kind of shout for connection.[4] It is almost as if the dog barks in part as a responsive reflex to our presence, shouting, *Look! A person! A wonderful, marvelous person!* Or, in another context, *Why am I alone? I need people near me!*[5]

In just a few short generations, the mind of these foxes had changed to the point where they could even respond to human gestures—*even though this was never an objec-*

tive of the breeding program. As Hare writes, "The experimental foxes understood human gestures, even though
the Russians had not bred them to be better at understanding human gestures. They were bred to be friendly
toward humans, and like floppy ears or curly tails, they
gained a better reading of human gestures by accident."[6]

Scientists are infatuated with the silver fox experiment because it seems to demonstrate a missing link,
showing not only how domestication happened with
dogs but also how quickly domestication can impact a
species. Interestingly, other approaches to domestication
have not led to the radical and expedient transformation
seen in the dog or the silver fox. It appears that domesticating for other traits, such as size or even usefulness,
doesn't deliver the kind of inner and outer changes that
breeding for the simple trait of tameness does.

For our conversation here, I see an important connection: the quality of tameness, or relational comfort
between animals and humans, opens the door to a more
complex emotion, one that could hardly be further from
a dog's lust for bacon.

Kind Speaks to Kind

We discovered earlier that empathy is the ability to feel
another's experience. It is fundamental to spirituality
because it takes us beyond ourselves into the encounter
with otherness—never for the sake of gain, but out of a
desire to be with and for the other.

Some theorists in the cognitive sciences believe that it was the tacit selection for kindness in our own evolution that allowed our ancestors to make the mammoth jump to *Homo sapiens*—which, as the species name *sapiens* indicates, means we are beings who are able to think about thinking. It is kindness that moves us to seek the minds of others, but even more, to allow others to connect meaningfully with us. Without kindness or empathy, there could be no collective mind, and without a collective mind, culture, religion, and art would be impossible.[7]

These scientists, then, believe that instead of seeking sexual partners that reinforced dominance, like other hominids and primates would, our ancestors favored partners who exhibited kindness. Selecting for kindness allowed the development of shared intentionality or empathy, which eventually lifted humans to an awareness of transcendent and spiritual realities. In kindness we find the seed of spiritual friendship, whether it is with our neighbor or with God.

John Cacioppo, in his book *Loneliness*, references a study that shows the power of kindness in making a happy marriage. Cacioppo says, "Sharing the joy in your partner's promotion, it seems, actually can be more important than being attentive when she gets passed over. Similarly, another study showed that when it comes to problem solving within a marriage, remaining cheerful and pleasant in outlook—even when that cheerfulness is combined with less than perfect communication skills—

was far more predictive of keeping your partner happy than was being a grump who somehow manages to do or say exactly the right thing."[8] Good listening skills, attentiveness, or problem solving in times of loss simply cannot measure up to the impact of cheerful sharing of each other's joy on the measure of people's happiness in marriage. We are wired to respond to kindness.

Here's what that has to do with dogs.

If early *Homo sapiens* communities were selecting for the traits of kindness, it wouldn't be surprising that they would be attracted to wolf-dogs that also showed kindness. Our ancestors took these kind wolf-dogs into their lives and chased away the unkind ones. Just as humans were choosing for kindness in their intimate partners, so they would also select the kindest puppies from the litters to sleep near and befriend their children. This echoed the spiritual reality of the shared life they were beginning to experience among themselves, and as the silver fox experiment shows, it would have taken a very short time for this selection of kindness to produce significant transformations in the wolves' bodies and minds, changing wolf-dogs into dogs.

If all this talk of kindness and empathy in a dog feels somehow weak or off point to you, trust me, it's not. The quality of kindness has the power to take the "wild" out of "wild animal" more than any other.

When we first brought Kirby home, Kara's granddad told us we were playing Russian roulette. Why would we risk having a wild animal under our roof, he would say, one that could eat the faces off our children while they were sleeping? When Owen was born, as much as I loved Kirby, and as kind as I knew him to be, I heard Granddad's words echo in my head. Any misgivings that might have lingered, though, evaporated one day when I saw kindness perfectly illustrated in the relationship between Kirby and my rowdy toddler. Owen had toddled over to Kirby and grabbed hold of his tail with both hands, like a sailor lining up to a rope. He gave a firm yank; Kirby gave a low warning growl. Owen pulled harder; Kirby growled louder. At this point, Kara leapt from the couch saying, "No, honey!" and tried to reach Owen before things went too far—but she was too late. In frustration, Kirby whipped his head around, with his teeth bared, and snapped his jaws closed *around his own tail,* yanking it out of Owen's little hands. Instead of a gut reaction to attack and destroy, our domesticated Labrador protected his toddler by biting his own tail instead of his boy.

Of course, sometimes enough is enough and a dog snaps; and like some people, some dogs have a short fuse; but when we stop to recognize how many dogs are living among us, it is amazing how kind and patient they are with humans, and how few stories there are of dogs returning the favor for meanness.

A Feeling for Feelings

New York can be the unkindest of cities, which may be the reason that 1.5 million pet dogs live in a place where it is an enormous hassle to have one.[9] Still, the hassle is worth the cost because humans need companionship like our lungs need air. In an urban environment, where many feel cut off from family and where social connections are hard to establish, we need the kindness of dogs all the more.

Yet, is what we experience as kindness and companionableness in a pet something close to genuine sympathy, even empathy?

Empathy is different from sympathy.[10] Where sympathy can compel you to say, "Man, it must suck to be you!" empathy does the more spiritual act of compelling you actually to *feel* another's need, joining his humanity and sharing in his experience.[11]

So can an animal without an extended consciousness, without the ability to think about thinking, truly have empathy for another dog, let alone a human? I base my answer mostly on personal experience, which shows me that there is a deep spiritual core in higher beings that operates outside of and beyond mere thought—and science is beginning to back that up.

In his book *The Feeling of What Happens*, neuroscientist Antonio Damasio tells the moving story of a patient named David who developed a severe brain condition that erased his memory. David could remember nothing from before his condition, nor anything that happened just hours earlier. When Damasio started meeting with David at his nursing home, it was clear that he particularly had no recall of the people he had known;

everyone he encountered was a stranger to David. This was a heartbreaking condition, for the human being needs others in order to live.[12]

Yet, Damasio found something profound in his time with David. While David had no ability to remember anyone, he nevertheless, without mistake, was able to respond to those who were harsh or kind to him. If a nurse had been mean to David, though he couldn't remember her or the incident, he would refuse to follow her and would protect himself, responding in a distant way toward her. And the opposite was also true; if a nurse had been kind to David, when encountering this nurse again, David would react with emotions of happiness and cooperation.

Damasio uses this story to show how feelings have a logic that goes beyond pure rationality, arguing that our minds are more than rational machines. For me, this story shows the spiritual depth of kindness—how it is hardwired deeply within us. The structure of David's brain kept him from the recall necessary to move a person from stranger to friend, and yet David mysteriously wasn't left without connection. Where his cognitive skills failed him, he could reach beyond cognition through a more primitive—or perhaps we should call it a more highly developed—intuitive capacity. Kindness seems so hardwired in us that even if our brains fail us, there is a backup system that can still send and receive it.[13]

In the same way, because our dogs have been formed from the time of early domestication by their capacity for kindness, they are able to *feel us* in a way that can rightly be called empathy.

Kara and I used to have a running joke that during every single episode of the show *Parenthood,* she would cry at least once. She would refer to it as the "Damn you, *Parenthood!*" moment. Whenever that moment came, Kirby would awaken from wherever he was napping, and come like a homing pigeon to place his front paws on her chair and press his nose toward her tears in concern. Kara would respond, "Kirby, I'm fine! It's just a stupid, awesome TV show." But Kirby couldn't tell the difference; all he knew was that she was sad, and he longed to share in it and comfort her.

A friend of mine tells the story about the hellish six months during which she lost her mother and her marriage. She sat for hours in the living room, staring into the distance, her grief so heavy she couldn't even make out how she actually felt. Yet, in a kind of pattern, every twenty minutes or so her dog would enter the room and sit next to her for a few minutes, as if feeling the tension and sadness within her. He'd lean in and look her in the eye, placing his chin on her arm. His touch would bring tears, and he would move his head to her breast. She would then pet him and feel better, feel understood. After a few minutes, seeming to be satisfied that he had achieved his purpose, he'd slide off the couch and leave the room—only to return twenty minutes later.

Popular dog writer Jeffrey Masson quotes Heini Hediger, the director of the Zoological Gardens of Zurich, who writes that "only the dog seems capable of reading our thoughts and 'reacting to our faintest changes of expression or mood.' German dog trainers use the term *Gefühlsinn* (a feeling for feelings) to talk about the fact that a dog can sense our moods."[14]

On this score, they save us from our own "higher" selves. Think how often we override our empathy reflex so that we can succeed in other ways: get to work on time, save money, protect ourselves and our families, beat the competition in contexts where kindheartedness will only take us out of the running. We can quite easily suppress, and eventually forget, the very depth of feeling that defines us as children of God.

But a good dog won't let us forget. *I like to be near you,* the dog will tell us in a hundred ways. *I think you're amazing.*

Or, *I can tell you're hurting, but I won't leave you.*

Or, like that happy spaniel at Gate A32, they somehow telegraph just the right message of reassurance to a roomful of strangers.

Empathy Matters

Canine compassion has been known to heal emotional wounds suffered by at-risk youth, prisoners, people suffering from illness, children, and more. Anthropologist Darcy Morey points to the dog's particular curative powers with the elderly. He says, "Based on interviews with over 900 elderly people in urban settings in southern

California, [the researcher] Siegel reported . . . that the frequency of medical doctor contacts was lowest overall among those who owned dogs, as opposed to those with other pets. In reviewing some of the then-current information on this subject more than a decade ago, [another researcher] Beck found that '[w]hile animal ownership generally had value, the most remarkable benefits to health were for those who own dogs.' "[15]

Robert Bierer, a researcher at the University of New Mexico, makes a similar point, showing how deeply a dog impacts a child, covering the child with a blanket of kindness and empathy. Bierer studied kids ages ten to twelve, examining the impact of dog ownership on children's social skills. "People have known for years that dogs are good medicine for children," he says. "What I found, is that preadolescent children with pet dogs have significantly higher self-esteem and empathy than children without dogs. These higher ratings in self-esteem and empathy hold true whether the dog is 'owned' individually by the child or by the entire family. That means that just having a dog in the house makes a difference, regardless of whether the family is headed by a single parent, the mother works outside the home, or the child has siblings."[16]

One day, shortly after Kirby's death, I found myself sobbing in front of my computer at a YouTube video. In the frame sat a small boy, four or five years of age, with Down syndrome. Next to him, watching him adoringly, protecting and playing with him, was a large yellow Lab.

The dog seemed completely attuned to the boy, wanting nothing more than to be fully with him. With every bounce and wag, it seemed to me, the dog was pleading with the boy to grasp a huge truth: *No matter how this world might treat you, I think you're amazing.*

In the background of the video, against the sound of traffic and city noise, I could hear the voice of the person filming the video. He, too, was part of the moment—drawn in, laughing in sheer pleasure at the dog's insistent attention to the boy, and the gentle yet absolutely tangible, kindness that held the two together. There I sat, a full-grown man, blubbering into the screen while a boy sat and a dog gave his kind, empathic attention.

It wasn't only because I'd just lost my own dog. There was more. I was seeing playing out before me that same mystical connection, that same clear message from a higher place that Kirby shared with my kids pretty much every day. This boy was experiencing the thing we long for most as humans, and watching YouTube, I got the message again, too. I thought it went something like this: *You are loved. You are mine. You are beautifully and wonderfully made.*

I was experiencing a sacred moment.

8

Bonding Fever

Three days after Kirby was put down, we went on a family vacation. None of us felt like going, but the cabin was already booked. It was our summer routine: every August, we loaded up our car with suitcases, water toys, and fishing poles and headed to the lakes in the Wisconsin Northwoods for a vacation with Kara's extended family. A major element of the fun had always been watching Kirby. He loved to swim, and there was all the space and time his tennis ball addiction needed for a real bender of a week. Plus, there was free food.

We loved Kirby, but he was a particular kind of villain to Kara's sisters. At least a few times over our week together, the whole crowd would pull out the grill, set up chairs, and cook out, eating down by the lake. Parents would load plates full of chips and hot dogs for the two-, three-, and four-year-olds who made up the family reunion. The adults would settle the kids in, food on their laps, leaving them happily taking first bites. Then, as soon as the big people's attention turned to their own food, Kirby would sneak up on the children like a silent ninja. Down the line of little laps he'd go, one by one, gobbling quickly off each kid's plate, inhaling Doritos and swallowing hot dogs whole. By the time the parents spotted him, Kirby would be a few children down the line, serenaded by a chorus of wails and tears, and Kara's sisters yelling, *"Kirby! No!"* This would only speed him up, though. Faster and faster he'd go, ketchup smearing, children screaming, chip crumbs sticking to his whiskers—until someone finally reached and yanked him away. Then he'd trot off looking pleased.

That wouldn't happen this year. This year, my kids cried for most of the three-hour drive, pleading that it wasn't right even to go there without Kirby. Once we arrived, we did our best to enjoy ourselves, but it often felt impossible. We saw Labradors everywhere: on sidewalk jaunts, at gas stations, on billboard advertising. We kept saying how much we wished we'd had one more summer trip with Kirby.

We noticed Kirby's absence most when we visited Kara's uncle Brent. Brent is a classic Northwoods brute. As ingenious as he is tough, he has made a very nice living for himself off many acres of woods he bought back in the late 1980s. He's used his land for real estate, farming, and raising deer (and even, at one point, emus).

What Brent enjoys most, though, is hunting, and his land offers ample opportunity. At least part of each of his days is spent driving his acreage in his four-wheeler, doing odd jobs and preparing his tree stands for hunting.

Yet, Brent never does these jobs alone. Before he can even get himself to the four-wheeler for his daily drive, his black Lab, Bear, is already sitting in the passenger seat. Bear lives to join Brent in his work. Every day, the tough woodsman and his black Lab cuddle chin to shoulder, like two 1950s teenagers in an old couch-bench Chevy, as they race over hills and across trails. They have quite a bond, and watching it made me miss Kirby. More than that, it gave me a name for what I missed: I'd lost a bond; a deep connection had been severed.

Robert Bellah has told us that there are three hard-wired experiences within us that lead to the spiritual. In the last chapter we saw how dogs draw out empathy in us (and experience it themselves), but how does our relationship with dogs meet our need for bonding? And what does bonding have to do with spirituality?

Clumsy Childhoods

Our need for bonding has its roots in the fact that humans have a much longer and therefore more dependent childhood than any other species. Donald Winnicott, one of the great psychoanalysts of the twentieth century, was famous for saying that there really is no such thing as a baby—only a baby and someone. For most of human history, our much shorter life spans meant that nearly all our lives were spent as either a child or a parent, and our close-knit communities meant that even if we were in the unusual period of *not* being parented or directly parenting, there were always children underfoot.

The bonding that has its origins in parenting is not optional. As parents, we may think we can opt out of parenting our child (which, tragically, happens all too often), but the child has no such choice. For the child to *be* at all, to follow Winnicott, she needs a long-term, primal attachment to another human. All of us have been, and remain, someone's child, and therefore we cannot escape the deep mark of attachment (or, in some cases, a lack of attachment) left on our being.

You could think of this dependence as the fatal weakness of our species—after all, it makes us needy, immobile, vulnerable, and clumsy during our formative years. Actually, though, it's our genius. The utter vulnerability of the human baby creates a deep spiritual reality of connection between the parent and child. Our own nurturing experience of attachment, of learning within days to share the mind of our mothers, prepares us to be in meaningful relationship with others, beaconing into the face-to-face encounters that deliver the spiritual. Our vulnerability is what leads us to select first and foremost for kindness.

You and I need bonding like we need air and water, and that's hardly an exaggeration. As psychologist John

Bowlby says, "The young child's hunger for his mother's love and presence is as great as his hunger for food, [and her absence produces] a powerful sense of loss and anger."[1] This hunger lessens as we grow to maturity, but it never entirely goes away. Human beings have a kind of spiritual heartbeat, spending the rest of our lives looking for things to attach to and bond with.

For other animals, it's a different story. As newborns, they imprint on caregivers, who help move them from utter dependence to the ability to meet their own needs for food and sex. For example, the bond between a mother ape and her grown offspring fades once the young ape can independently seek its own needs. That's because once the maturing process begins, bonding is not only useless but also a detriment to the offspring's survival. Once other mammals are able to feed themselves, the pull of attachment ends, broken like a fever, until the fever returns again when the animal must care for its own offspring.

Humans need attachment not only to get started but also throughout our lives. However independent you and I might think we are, the truth is we can live and flourish only inside the gravitational pull of meaningful bonds with other persons.[2]

Yet how, you may be wondering, does all this apply to our canine housemates or, more specifically, to the spiritual connections between us?

Without Humans, There Would Be No Dogs

Drawing from Winnicott, we could say that there is no such thing as a dog, only a dog *and someone*. If there were no human beings, there would, in all possibility, continue to be wolves, but without human beings there would be no dogs. The dog, like the human being, must be understood always in relationship to humans. It survives because it is bonded to us; by definition, it exists because of us. Though we can imagine a lone wolf, and there are other animals that spend the greater part of their life alone, there is no such thing as, and could not be, a lone dog. Bradshaw explains:

> The dog's sociability is even more remarkable when compared to that of its ancestors. Wolves from different packs try to avoid one another; if they do meet, they almost always fight, sometimes to the death. This is not unusual. Modern biologists view all cooperative behavior as exceptional, because the default behavior of every animal should be to defend itself and its essential resources—its food, its access to mates, its territory—against all others.[3]

Dogs, however, like humans, feel the need for attachment their whole lives. Their "default behavior" is to stick close to their people, choosing to bond with us, a different species, even over other canines.

In the late 1990s, when scientists Adam Miklosi and

Josef Topal wanted to test relational bonding in dogs, they copied an attachment experiment done on babies in the 1970s. In the original experiment, called the "strange situation test," a one-year-old child was placed with his or her primary caregiver (Mom) in a roomful of toys and invited to play with them. Right away, the researchers noticed that children with strong attachment kept returning to their mothers as a kind of home base as they explored and played.

After a time, the caregiver would depart and either a stranger would enter or the child would be left alone. The research team recorded how the child behaved in these new situations. What gave them the clearest picture of attachment, however, was how the child responded when the primary caregiver returned. Those with secure attachment celebrated their mom's reappearance with kisses and hugs. Those with poor attachment ignored Mom, or hit her in anger for having left.

How would dogs fare in their strange situation? Miklosi and Topal wondered.

To find out, they set up parallel scenarios. As its primary caregiver watched, the dog was invited to play and explore in a roomful of toys. After a period, the dog owner would leave and a stranger would enter. What happened then?

To Miklosi and Topal's astonishment, the dogs responded much like the children with secure attachments, playing and exploring confidently, and checking in with "Mom" regularly. When the caregiver left, the dogs became more cautious, playing and exploring less, until their owners returned. During their owners' absence, some dogs ignored the strangers and even the toys and sat down to wait. Others began to bark or scratch at the door.

If you own a dog, you *know* how the dogs in the strange situation test responded when "Mom" returned. Like well-loved kids everywhere, the dogs with strong attachment exuberantly welcomed their returning owners.[4]

All this leads me to believe that my dog skeptic theologian friend hadn't noticed an important dimension in his relationship to his dog. Yes, dogs want food from us, but all things being equal, they don't want it from just anyone. They want it from *their* person.[5] Those big baby eyes communicate and invite a real sense of attachment that's targeted toward a particular person. Masson says it poignantly: "Love on the part of the dog does not seem conditioned merely by what we provide the dog, nor simply a recognition that we are a source of food. A dog does not love a robot that gives it food, but is capable of loving people who never feed it."[6]

Hare adds another interesting layer: "Not only do dogs prefer to spend time with a human than with one of their own species, but they are so focused on us that sometimes it can work to their disadvantage. For example, we know how much dogs love their food, and if they have to choose between a small pile of food and a large pile, usually they choose the larger pile. But if your dog sees you repeatedly choose the small amount of food, they are more likely to choose the small amount."[7] It is a remarkable creature that does this! Our dogs are so deeply shaped by their bonds to us that they are willing to follow our lead, even if it costs them personally.

This was illustrated to me recently when a friend told me about his experience of putting to sleep Edmund, the family's beloved Chihuahua-terrier mix. The rapid spread of cancer throughout his small body had taken everyone by surprise. In the vets' office, my friend and his wife were saying their good-byes.

"We were in shock, honestly," the friend told me. "In just days, it seemed, Eddie had gone from the high-energy light of our lives to a frail shadow of his former self. Everything about what had happened, and what was about to happen, seemed terribly wrong."

As his weeping wife held their dog, Edmund was still determined to give back. Even in his pain and weakness, he struggled with single-minded purpose to stand in her lap, and then he reached up to gently, patiently lick the tears from her face.

The New Work of Dogs

Before we had children, when Kirby was just a puppy and still able to be carried in our arms like a baby, we took him on a Saturday afternoon adventure to PetSmart. A nice young salesclerk was doing a Kong demonstration, showing us how you could fill the beehive-shaped rubber toy with almost any food, and how much dogs loved it. She shoved a Kong filled with peanut butter into Kirby's face, and he took his first excited licks of peanut butter. To my surprise, Kara was livid. I could see, for the first time in our marriage, the intense protective emotions of motherhood seeping from her every pore. When we walked away, she fumed, "How could she do that? Kirby has never tasted peanut butter before, and she didn't even ask if he could have any! She didn't even ask me if it was okay to give him some! *I* wanted to be the first one to give him peanut butter, not some stupid girl at PetSmart!"

We chuckle about it now, but Kara was in her late twenties at the time, an age at which, even a hundred years ago, she would likely have already had multiple children. She was projecting her parental bonding onto Kirby, and even she was surprised by the power of this feeling. After Owen was born, her bond with Kirby seemed to lessen, but as Owen began to grow, he began to exercise his own needs for attachment beyond Kara (and me) by bonding intensely with Kirby. Kirby came to represent safety, assurance, a sense of home. Owen missed Kirby when we were gone, and sought him out the moment we neared home.

In a world that so often divides us more than bringing us together, that keeps us moving more than keeping us deeply rooted, this need for attachment suggests what is becoming the "new work of dogs," to borrow a phrase from the journalist Jon Katz.[8] Katz believes that dogs, in both beautiful and peculiar ways, have shifted from partners in work to something more emotional and familial.[9]

Because our cultural realities often prevent us from experiencing family and parenting for long stretches of time, our dogs have become more like our children.[10] We project our need to parent onto our dogs. We call ourselves "Mommy" and "Daddy," bundling our dogs into sweaters and booties, talking to them in that incessant baby talk that so annoys our friends. In later years, we find that having a dog can help make up for the absence of our children after they grow up and move away. We no longer buy dogs; we adopt or foster them. Yet these terms say less about the dog's experience, I believe, than about what we want our dogs to give back, which is to provide for our own needs for family, intimacy, companionship, and commitment. Dogs come to embody a sense of family and home, and our attachment to them can transform how we feel about a place or time in our lives.

Interestingly, we live in a culture that often refuses to believe there's something ingrained in our being that yearns to parent. We're willing to admit that other animals go through times of being in heat, driven by the biological need to mate, birth, and parent. But not us! And yet we carry our dogs in purses on planes, dress them in raincoats, and say things to them like "Do you want Daddy to take you on a walk? Do you, my sweet baby?"

Fortunately, our dogs happily play along.

More and more, the new work of the dog may be to give what our frantic, fast-paced, uprooted lives have taken away. Still, I see its role as being in deep continuity with what the dog has always done for our species. It helps us *be* and *become* more fully human by reminding us that what we are made for, and long for, and have to give, is connection. Through caring for and being cared for by a dog, we encounter a recurring spiritual experience of bonding, a beckoning to come out from the shell of our self-interest and see another, to respond to its need, and to experience the joy of our being together. Our bonding to our dogs and their bonding to us show a deep relational connection that sends a current across our hardwired spirituality.

And, what's so fun, is that often, dogs do this through the simple, and surprisingly spiritual, act of playing.

9

Doggy Play as Soul Talk

It's a strange technological marvel (and maybe even a uniquely modern way of grieving) that allows us to sit in front of a computer and watch high-quality videos of moments from our past. After we lost Kirby, Maisy would ask to look at old pictures of him and especially videos. So we would sit together on the couch with my laptop, reviewing Kirby's life with us.

We watched him leaping around in the snow and swimming in the lake. We giggled at a scene of him stretched out on the living room floor while baby Maisy crawled toward him and then curled her body up against his. We laughed at a clip from a birthday party where the camera caught him table-surfing a piece of cake. Then we came across a clip of a family game I had forgotten about.

For several weeks, it had been three-year-old Owen's

favorite pastime. He called it, in his high-pitched squeak, Play-Hide-and-Seek-and-Give-Kooby-Bread! All one squeak. In Play-Hide-and-Seek-and-Give-Kooby-Bread, Owen would climb a kitchen stool to stand at the counter and tear a piece of bread into small bits while Kirby watched. When Owen had the bread broken and arranged in a pile at the edge of the counter, he would stand at his command post on the second step of the stool, then point with great authority at the floor below him and declare, "Sit, Kooby!"

The dog, easily three times Owen's size, would obey. Mind you, Kirby could easily have reached up and gobbled the bread, but, no. He sat and waited and drooled.

Next, Owen would tell him loudly, "Stay, Kooby!" and clambering laboriously down from his perch with bread in hand, he would toddle off. When he was in some corner of another room and well out of sight, he would hide the bread. Then he'd come running back with barely contained giggles and shout, "Go get it, Kooby!"

Only then would Kirby race to discover the hidden treat. Over and over, the two would play their game—the big dog, all wags, eager eyes, and rapt obedience; and the little boy, giddy with happiness in his newfound power. A similar scene happens every day in parks and living rooms and front yards. Common, we all agree, but when you think about it, absolutely astonishing.

Consider carefully what Maisy and I were watching played out on that screen: in our kitchen, two mammals of very different species, one virtually a baby and the other an adult, learned to communicate clearly and work cooperatively to create a complicated experience. This shared experience depended on delayed gratification, but it delivered such delight for both that they were happy to repeat it, and repeat it again.

Why? To have fun, that's why. To play.

What can family scenes like these tell us about dogs and humans? And what might play reveal about our natures, and the mysterious interconnectedness between our dogs and us?

Play is the third in Robert Bellah's triad of experiences that hardwire us for the spiritual. We've seen already that not only are experiences such as empathy and bonding hardwiring us for spirituality, but dogs, too, seem to relate to the spiritual hardwiring within us. As we'll see, though, there really can be no bonding or empathy without play.

Playing to Win

The first thing we can say with certainty about Kirby and Owen's kitchen game is that, in the great diorama of birds and beasts in our world, this kind of cooperation is nothing less than an oddity and a wonder. To show you what I mean, let's take a look around.

Most mammals (and some birds) play when they are young, but in these cases, play is mostly serious business. Playing for young wolves and apes seems to be a kind of indoctrination into the organization of dominance and submission, exposing like a neighborhood backyard football game the beginnings of the script that says who is weak or shy, and whom not to mess with. The young are allowed to tumble and wrestle with one another and even bite or pull on their elders, but in virtually all cases in the animal world, direct engagement in play lessens when the animals come of age.

Humans, too, use the physical push and pull of play to test ourselves and others, to learn how our strength of mind and body matches up with that of our play-mates, and to teach us what we can accomplish only in teams. The rough-and-tumble identifies winners and losers, yes, but it also leads to a sense of bonding and respect. This explains why football players who, just minutes after trying to crush one another, will shake hands once the game is over, patting each other on the back and exclaiming, "I love you, man—great game!"

Yet play between pets and people is play at a whole other level: it represents a complex, lifelong, interspecies relationship, and there really is no corollary for that in the animal world. In fact, what play teaches about domi-nance and submission, individual strength and team possibilities, points to exactly why animals tend *not* to play outside their own species. If play is about work-ing out dominance and submission, or forming deeper bonds, why would a bird ever play with a monkey?

This brings us back to what's so special about Owen and Kirby's kitchen game. "Play between dog and owner is such an everyday occurrence," writes Bradshaw, "that it's easy to lose sight of the fact that interspecies play is otherwise very rare (indeed, virtually unknown outside the realm of domestic pets). To be successful, play requires well-synchronized communication; both partners must be able to convey their intentions precisely while at the same time convincing each other that they're not using the game as a prelude to something more serious, such as an actual attack."[1]

Sure, Kirby liked his bacon (or in this case, bread), but the game for him and for Owen was about more than food, and more than acting out roles of dominance and submission. So what compelled them to play in the first place?

I think it had more to do with bliss. The famous Harvard psychiatrist George Vaillant noted that as we age, humans give up rough-and-tumble play for sport, singing, and especially dancing, and how these shared experiences are so often "accompanied by both joy and happiness."[2]

A boy and a dog reveling in joy and happiness? Yes. That's much closer, I think, to what made Owen squeal, and Kirby wait and drool and race to the hunt, and the rest of us watch with such pleasure.

Playing for Pleasure

Playing for pleasure is central, I believe, to our spiritual nature as humans. In this kind of play, we simply enjoy our bodies and minds in motion.

Imagine a mother with her months-old baby. You know the scene: The mother lies next to her child. They're gazing at each other, enjoying face time. She touches his nose. He coos and squirms. She pretends to hide, she sticks out her tongue, she tickles his cheek. Together, they smile and babble at each other.[3] There is no winner or loser in these activities, and mother and baby have no objective except to encounter and enjoy each other, and to share more fully each other's worlds. It is a form of play that communicates to the child that he is human, and invites him through play to taste the joy and mystery of connection.

Unlike most animals, humans never outgrow the desire for this kind of play. We tease our coworkers, loll around in boats, sing at the top of our lungs at rock concerts, and keep running jokes going for years. Through these happy rituals, we enter deeply into the lives of others and express our continuing openness to the world as a good and welcoming place. We embrace beauty as our bodies and mind find harmony with other bodies and minds.

Maisy laments deeply on summer days when there is no one to play with. On the other hand, when a playmate decides to use their play for dominance, she storms back inside. What is going on here? Like every child, she is looking for friendship and connection, and play is a primary means to that end. So when a playmate turns a shared experience into a test of dominance—bullying, she would call it—Maisy perceives it as a great violation. The only contingent factor in play for her is that the other person participate in a way that allows her and her friends to share intentions and experience kindness, with no ulterior motives. She wants someone to reflect back to her that she exists, that she matters, that she is worth spending time with, and to serve as that kind of friend for another. Play creates a personal connection that brings with it valuable emotional payoffs.

Yet there's even more at stake, I believe. The play of kindness and bonding can take us out of ourselves and actually attune us to the transcendent. Brian Sutton-Smith, in his book *The Ambiguity of Play*, explains: "Children's play fantasies are not meant only to replicate the world, nor to be only its therapy; they are meant to fabricate another world that lives alongside the first one and carries on its own kind of life, a life often much more emotionally vivid than mundane reality."[4]

Robert Bellah says that humans have benefited for at least the past quarter of a million years from "very extended adult protection and care of children." A particularly important result, he writes, is that "children can spend a lot of time in play—to think of the universe, to see the largest world one is capable of imagining, as personal."[5]

If you see the possibilities here, you'll understand why the play of kindness and mutual cooperation—with each other and with our beloved dogs—can be seen as a kind of spiritual workshop. It excavates and shapes our soul to rise higher, to experience deeper, and to receive more from the God in whom, as Paul said, we live and move and have our being.

"Jackson, Are You Ready for Your Bath?"

A few years ago, I came across a YouTube compilation of dog videos from an episode of *America's Funniest Home Videos*. My favorite clip showed Jackson, a gangly young retriever, standing in a bedroom doorway. The woman behind the camera asks, "Jackson, are you ready for your bath?" an invitation that sends the dog racing to dive under the bed. A moment later she asks, "Do you want to go for a walk?" and he comes shooting joyfully out from under the bed, tail wagging in anticipation, and sits down in front of her with wide happy eyes, panting in joy. She asks again, "Would you like to take a bath?" which sends him plunging back under the bed. The scene ends with her saying, "Come on, let's go for a walk!" Out flies Jackson from his hiding place, all eagerness and joy, ready to go.

Clearly, Jackson understands the different meanings of the words *bath* and *walk*. The thought of taking a walk with his owner is so irresistible, that he is willing to risk coming out every time it's mentioned. Like Charlie Brown with Lucy's football, he knows it may not happen, but *this time it might*. Though unlike Charlie Brown, Jackson doesn't reflect on his odds. He doesn't hem and haw and wonder if this walk will become a bath and whether it's better to stay put. His drive to play is so strong, so wound into his very being, that he pops up alert and ready each and every time.

Interestingly, when given a choice between playing with other dogs and playing with people, dogs will choose people. Dogs will even change their playing behavior to make a game more people-friendly. For example, when playing tug-of-war with another dog, the game is clearly all about competition, and both dogs will compete until there is a winner. But when a person is involved, dogs will regularly *choose to lose* in games of tug-of-war with the hope that it will keep the game going. Dogs will tug and tug, and then let the human win, because they are sensitive to our feelings. They don't want us to get frustrated and quit.

Creatures that need bonding throughout their whole lives engage in play as a way of entering genuinely into the mind of another. Like us, a dog never outgrows the capacity for having fun, and, unique among animals, the dog is consistently able to play across species. So a dog like Jackson could befriend a chimp, a bird, an elephant, or another dog, while at the same time loyally and exuberantly accepting every invitation for a walk with his owner.

Downward Dog

Maybe you've noticed: dogs often start play sessions with a bow, stretching out and down on their front legs in a way that gives the Downward Dog pose in yoga its name.

It's a ritual of sorts that invites another dog, or a person, to come and join the game that has no scoreboard. Bradshaw explains:

> When two dogs are playing together, play-bows are much more likely to occur when they are facing one another than when one is facing away, indicating that dogs are sensitive to whether or not their play-partner is paying attention to what they're trying to convey. Dogs that want to perform play-bows but are being ignored have a variety of ways of getting another dog's attention, including nipping, pawing, barking, nosing, and bumping.[6]

When I see a dog bowing in play, I'm reminded that doggy play with humans is actually part of a more meaningful act, a spiritual one that ties us more closely to each other and that leads us deeper into mystery and wonder, maybe even to touching the bliss of God.

When Kirby was an overly energetic puppy of six months, he had a way of getting himself in trouble as he fought to feed his insatiable desire to play. His most notorious moment was one morning when Kara had just gotten out of the shower. Wrapped in a towel, she opened the bathroom door to find Kirby standing there, still as a statue, eyes on her face, his body crouched in a half bow, enticing her to play. He also held a paring knife in his mouth by the handle, blade exposed like a mugger demanding her purse.

Kara screamed. "No, Kirby!" she pleaded. "Drop it!"

Unfortunately, to Kirby, her scream was a sign that the game was on. Down he went into his bow, waving his tail in glee, and the chase was on.

For ten glorious minutes, Kirby led Kara in a high-speed scramble. They looped through the kitchen and back through the living room, then back through the kitchen again, around and around, all the while with the blade in Kirby's mouth slicing the air. Kara's exasperated screaming turned to swearing, but to Kirby, this, too, seemed part of the game. He'd stop and challenge her to approach, but just as she closed in, he'd dodge and reroute. She didn't give up, though. Dripping with water and frantically trying to keep her towel in place, she chased and yelled and begged him to stop.

Only when Kara had barricaded the kitchen doorway with overturned chairs was she able to corner Kirby and retrieve the knife from his mouth. By then, though, Kirby had gotten what he wanted. Practically sighing with satisfaction, he licked her face as though to say, *I know. That was fun, wasn't it? You're welcome.*

10

A Healing Presence

When my friend was picking up his new puppy from a breeder, the breeder asked him where the dog would sleep. My friend explained that they had a little kennel ready for the dog.

"Okay," she said, "that's good for when the puppy is little, but how about when it grows? I mean, will the dog sleep outside or in the house?"

The puppy was going to be a family dog, he said, so he imagined that once potty training was accomplished, the dog would sleep beside their bed.

"Or maybe *in* it!" the breeder exclaimed. "A dog in your bed is one of the best parts of life! There's nothing more relaxing than falling asleep with your dog curled up against your body, or cuddling in bed together on a lazy Saturday morning. I love it. For me, that's total peace and rest."

Dog in the bed or dog somewhere else? This is a question every couple must decide. Kara and I settled on dog on the floor, but in the mornings, Kirby would sneak into our bed and wriggle his soft, black body in between Kara and me. Before we had kids, he was our family and our reminder that family always starts close up and personal. The clock could be ignored. The stuff that had to get done could wait. To begin a day right, we first had to do some serious snuggling together.

That priority gets quickly forgotten in the frenzy of our days, I've found. We live in a 24/7 world, plugged in, powered up, and accessible night and day, where the boundaries between work and home life have become hopelessly blurred. We forget how to slow down, how to be fully *here* before we rush off to *there,* how to master *being* before *doing.*

Dogs have no such problem. Dogs force us to stop and put down work. They need to be walked, brushed, and cuddled, and their need is grace to us, demanding that we stop and tend to our own souls as we tend to their needs. Dogs don't care if we made VP yet, checked off all the items on our to-do list, reached our ideal weight, or cleaned the garage. They give us permission to rest. They recognize our worth simply because we exist and we're their humans. They tell us to enjoy life as it is, not as we are working for it to become.

What a remedy for the soul this can be!

Sometimes, when I felt super busy and stressed out, I would find Kirby conked out on the couch, legs splayed, not a care in the world, and I would feel a pang of jealousy. Why couldn't I be that free? Why couldn't I live in the moment, like a dog napping in a sunbeam?

Then I came to realize that this reminder was itself a gift. By his example, Kirby could draw me back—if I let him—to a simpler, better quality of life. Nearly every form of spirituality or religion teaches that we must be attentive to the now. And here was Kirby, my live-in reminder to live gratefully in the present.

So, really, I see it as a choice. We can think of our dog as a beast that makes constant, primitive demands and sprawls over the couch we worked hard to pay for. Or we can see him as a natural-born spiritual director and healer, one who excels at helping us receive the eternal that's waiting in each moment.

Comfort in Trauma

I've always been fascinated with the capacity of dogs to sneak into our ordinary days and show us, in the words of Jesus, that the Kingdom of God is already here and among us. I'm not saying this is the exclusive ministry of a dog, mind you. A sunset over your favorite lake or a deep conversation with a friend can do that, too, as can sitting with a dead bird as it dies, or hearing the sound of your baby's laugh, or, if you're lucky enough to visit Florence in your lifetime, the majestic beaconing of the Duomo.

What dogs might do best, though, is show us God in

the very worst of times. Take these opening lines from a *New York Times* article in June of 2016:

> On the Monday following the Orlando massacre, 12 golden retrievers arrived in the Florida city. They had come to offer comfort to some of the victims of the attack, the families of those killed and the emergency medical workers, as well as anyone else in the city in need of some canine affection after the deadliest shooting in American history.[1]

The highly trained animals were part of the K-9 Comfort Dogs team based in Northbrook, Illinois, and this wasn't their first outing. They had brought consolation to victims after the 2013 Boston Marathon bombing, too, and to the aftermath of the 2012 Sandy Hook shooting before that. The Comfort Dogs excel at providing a feeling of safety, at inviting people to let down their guard and express their vulnerability in the midst of traumatic circumstances.

"Dogs show unconditional love," a program spokesperson told the paper. "We've had a lot of people here that start petting the dog, and they break out crying."

The truth is, for centuries dogs have brought solace to humans going through hell. The thirteenth-century saint Roch, of Montpellier, France, is known as the patron saint of dogs. His story shows why. Roch was a wealthy young man born with a red birthmark on his chest in the shape of a cross. After the death of his mother, he feared that he had not been living in a way that honored the mark of the cross on his body, which he understood as a calling to serve God fully. So he sold all he had, gave the proceeds to the poor, and went on a pilgrimage. According to some accounts, all he took with him was the family dog.

At the time of Roch's pilgrimage, the bubonic plague was ravaging Europe, reducing the populations of major cities such as Florence by up to a third. It must have felt like the Apocalypse. Yet, in these awful years, Roch took up a ministry to those who were routinely abandoned as soon as the plague struck and left to die alone.

Where others shrank back, Roch and his dog would approach the sick without fear. The dog would lie down next to sufferers and lick the sores that often accompanied the disease. Roch saw that the dog's patients improved; some even healed completely, and even those who remained sick nonetheless seemed to recover their humanity. Watching the dog's ways with the sick inspired Roch. To him, the dog was a witness to these beleaguered communities that even the very ill among them were still human, still deserving of recognition.

When Roch himself came down with the plague, he retreated to a small hunter's shed, expecting to die alone, but his dog refused to leave his side, and tended to him until he recovered. After he was healed, Roch and his dog continued to minister to others.

In times when evil and suffering seem to prevail, an animal without the gift of speech can remind us of the truth we need but can no longer recall. Thomas Merton's words distill this well: "We are living in a world that is absolutely transparent and the divine is shining through it all the time."

The Paws of Peace

As we saw in the snuggle scene at the top of this chapter, the healing gifts often come on a smaller scale, and in quieter moments, too. Yet this doesn't make them any less important.

As a child, I struggled with reading, especially reading aloud. I know well the paralysis that comes with performance anxiety: the mind goes blank, the words get stuck, and nothing comes out. If I was worried what the teacher thought, how I ranked, how I was doing, what my friends were thinking, I was cooked. Everything I had learned about how to read would fail me. What would have helped me, but what I unfortunately never had, was the ministry of a reading dog.

Today, elementary schools and libraries across the country have programs for kids who are like I was. For example, a few days a week at schools in the Northeast, leaders from the Good Dog Foundation bring dogs to the school library. A child sits down on the floor next to a dog and reads aloud from a book. There is something magical about it. After lying next to Pepper, a slightly overweight border collie, and reading him a book, seven-year-old Jessicah, who has always hated reading, says to the volunteer, "[He] loves when I tell him stories. I think he likes stories about turtles best, and so do I. He's the coolest dog in the whole world."[2]

I can vividly imagine what it would be like to be the child in that situation. To read to a dog whose big eyes took me in with simple pleasure, who laid her head on my lap with absolute ease to listen to my voice, would have made for an entirely different experience. The dog would have exuded patience, unconditional acceptance, and peace. The words I botched terribly would have captivated the dog every bit as much as the ones read perfectly. In that one-on-one relationship, the anxiety, self-doubt, and panic I used to feel about not being able to *do* something would have faded. I would have become free simply to be.

That's the power of a dog's attention. It moves us out of powerlessness, granting us the clear sense that we matter, that our timeline is the right one, that everything is going to be just fine. Barbara Christy, a third-grade teacher and facilitator of the reading club, concurs, saying, "Kids who used to slink into a room are walking tall, with their shoulders back and head[s] up. One young man used to stutter; now his speech is nearly stutter-free. I tell everyone there's magic occurring every week in this classroom."[3]

I'll admit, the theologian in me sees at work in this little scenario not magic but the ministry of God. Whether it's Kirby on my couch or a reading dog with an insecure kid in a school library, in both scenarios I see a powerful invitation to experience Sabbath. Sabbath is a ritual where we put down all our striving and simply rest, by God's invitation, in God's promise and provision. It is a time for healing. It reminds us that no matter what the world says about us, or what we say about ourselves, only another, higher power has the final say—that, at least for this moment, all is well.

11

Could It Be Love
They're Feeling?

Videos of war veterans reuniting with their families might be one of the defining Internet phenomena of recent years. Even if you haven't seen one, you can picture the scene. A soccer field, school assembly, or front yard serves as the backdrop for a cluster of kids who are unaware of what's about to happen. Then, the children turn around to see their mother in uniform approaching; or someone opens a giant box, and out pops their dad. The kids scream, cry, and laugh as they launch themselves into their parents' arms. Who can resist the poignancy of these family moments, or watch them unfold without tearing up?

Interestingly, another kind of veteran reunion video often outpaces the parent-child ones in online shares. It's the soldier-dog reunion. In these videos, the returning serviceperson walks into the room or climbs out of a car, taking her dog by surprise. The dog sees his owner from across the parking lot and bolts out of the back of a pickup to say hello. In their own doggy way, the canines laugh, cry, and hug for the camera just like any human family member would, rapturous with joy at the reunion.

These scenes prompt a question: We know children love their parents and feel their absence deeply, and vice versa. Dogs, too, clearly recognize and miss their owners—but what, exactly, do dogs *feel* at these times? To us, it looks like love, but could that be what it really is?

The Look of Dog Love

To answer the love question, we need to reach back over our preceding discussions about empathy, bonding, play, and reading human gestures—all the way back to my skeptical theologian friend. He put the question in simple, behavioral terms: Do dogs only act like they love us because it helps them get what they want? Or could it be that there truly is something deeper going on in there? And what *is* love, anyway?

In the Christian tradition, love is a profound feeling that expresses itself generously, even sacrificially, on the behalf of another. Love is what compelled God to send Jesus to the cross, turning death into life. Love for enemies leads us to pray for and serve those who don't return the favor. A mother's love for her son is why she's willing to sit in a freezing ice rink for what seems like the entire prime of her life, learning to love ice hockey. Her boy is crazy about hockey, and she's crazy about him, so she bundles up and steps into the icebox just to make him feel special.

When empathy, intimacy, and pleasure with another person motivate intentional acts of caring, that's love—and there's life-altering power in it. This is why in the New Testament, when Jesus asks Peter if he loves him, and Peter says "Yes," Jesus replies, "Then feed my sheep" (John 21:17). To love Jesus is to join him in his intention to love the world. "You cannot say you love God," Jesus says, "and hate your brother or sister," for you have not participated in the intentions of God to forgive, heal, and reconcile all that is broken (1 John 4:20).

Yet, wait a minute. We're talking canines here, not *The Greatest Story Ever Told*. We have to keep all four feet on the ground.

Still, we've been accumulating solid evidence in the previous chapters that dogs do indeed experience something that looks a lot like love. Here's what I mean:

First, we looked at a dog's *desire to be near* its human owner. Then we described a dog's astonishing and unique *impulse to know the intention, watch the face, and read the gestures* of people, especially the central person or people in its life.

From there, we took an extended journey into the building blocks of any higher relationship: *empathy* and *kindness, bonding* and *healing play*. We saw how dogs and humans give and receive these spiritually charged gifts in a well-defined, sharing friendship that's found nowhere else in nature.

Finally, we saw how a dog's *unconditional affection,* often its mere presence, can help humans calm down (so we can read for the teacher) and open up to the divine miracle of the moment.

All that doesn't exactly equal love, but in any human relationship, it would certainly be taken as that—and it moves us further from "dogs love bacon" and closer to "dogs love us."

The point is that love is a complicated thing; we both demand that those we love speak it out loud and claim that love is more than words. Still, few couples can actually move forward in their relationship until someone says those dreaded and glorious words, "I love you." Oddly, as much as we need to hear those words, we know that love is more than words, as the '90s band Extreme taught us with long hair, good looks, and the strums of an acoustic guitar.

More than words is all you have to do to make it real
Then you wouldn't have to say that you love me
Cause I'd already know.

So for human beings, love is about verbal proclamation and yet, equally, about actions and intentions. Clearly we know our dogs can't talk (though that would be awesome), but they do seem oddly wired for our communication. Also, though they can't directly talk, their intentionality seems strong enough for us to explore whether our butt-licking dogs do indeed love us.

Words for the Heart, and the Brain

Once, after Kirby stayed with friends while we were on vacation, our friend who had watched him said, "I've never seen a dog who knows so many words. It was almost eerie how he picked up on what we were saying." I wasn't surprised. Even before baby Maisy could call out to Kirby with "Come, Daaga!" when all she could manage was a delighted squeal or scream of summons from another room, Kirby would sense her intentions and come running. He was not only good at knowing words but he was also eager to use those words, or even nascent stabs at communication, to be with us.

Dogs are willing, even eager, to decipher and understand what our sounds mean. They will pretty much learn as many words as we take the time to teach, often amassing a vocabulary of one hundred words or more. (You can spell it out—*w-a-l-k*—or start saying "ambulate," so you and your spouse can make a plan without sending Hector the hound into a froth of anticipation, but since the new word is attached to one of Hector's deepest desires, you're probably just buying time.)

We take for granted our pooches' responsiveness to words, but we shouldn't. It is deeply meaningful to us that dogs respond to our words, because language is our primary means for connecting with others. It is harder for us to know if our cats care because they seldom come when we call them by name—and even if they do, they rarely come with the joy and excitement that our dogs do.[1]

This desire to respond to our words bears the weight of love because dogs sense the emotions behind our words. Dogs don't understand the meaning of every sentence, but they read our joy, sadness, excitement, worry, or anger with uncanny skill. And their faces communicate that our words have emotional impact on them. As I've just said, ask any boyfriend or girlfriend: words of love can feel an awful lot like the real thing.

Of course, for a more objective understanding—and one that is more likely to convince the skeptics—we still need to take a look inside, getting a little science to show that indeed this intention on the part of dogs to be with us by soaking up our words means love. Fortunately, in the last few years, scientists have done just that.

A few years after Hare and Tomasello's pioneering work at Emory, a neuroeconomist (yes, that's a real thing) named Gregory Berns set out to examine whether dogs had the mind to love people. To find his answer, he decided to scan dogs' brains to see if he could determine if we were indeed on their minds.

Taking an MRI of a dog brain isn't difficult: just sedate the dog and let the machine click and clack away. Yet this kind of scan would tell Berns and his team very little, for it would show a dog's brain only in neutral. What they hoped to show was a dog's brain in drive—the drive to respond to and be with us. For this kind of research, the functional MRI (fMRI) was the way to go—but for the researchers to collect the data, the dog would need to be awake. So, over several months, Berns trained family dogs (including his own) to do an extraordinary task: crawl into the noisy, enclosed metal tube of the scanning machine and lie completely still. This is something most adult humans balk at.

As the dog in the study lay in the tube, the family member who trained it would speak to it, giving it directions and the reward of human presence—and at strategic moments, a little hot dog. Sure enough, the images Berns was able to take were astonishing.

The first thing Berns noticed was that the dog's mirror neurons fired in response to the human handler. Mirror neurons are the neurons of cooperation, the neurological building blocks of empathy. Berns said, "Seeing this kind of mirror neuron activity in Callie and McKenzie [two dogs studied] meant that the whole dog-human relationship was not just a scam. If dogs had the ability to transform human actions into their own doggy equivalent, then maybe they really did feel what we feel. At least a dog version of it."[2]

Berns believed that he had shown that dogs have a mind for us, as Hare's experiment had also indicated, but did the fMRI scans show whether dogs love us as much as they love hot dogs? Berns concluded that they did. He says, "The inferior temporal activation told us that the dogs remembered their human family . . . These patterns of brain activation looked strikingly similar to those observed when humans are shown pictures of people they love."[3]

Another scientist, ethologist Vilmos Csányi, came to a similar conclusion about the dog-human connection. He wrote:

> I'm not a religious man, and I pause before using the word *soul*. But my experiences with the dogs in my life . . . convince me that there is some profound essence, something about being a dog, which corresponds to our notion of . . . soul, the core of our being that makes us most human. In human animals, this core, I am convinced, has to do with our ability to reach out and help a member of another species, to devote our energy to the welfare of that species, even when we do not stand to benefit from the other—in short, to love the other for its own sake. If any species on earth shares this miraculous ability with us. . . . it is the dog, for the dog truly loves us sometimes beyond expectation, beyond measure, beyond what we deserve, more, indeed than we love ourselves.[4]

What can we conclude? Well, since humans (thankfully) have never been able to reduce love to a particle or potion, we shouldn't expect to finally prove canine love with a brain scan. And let's admit that in our human relationships, you and I carry proofs of love in our hearts, not file folders.

Kirby didn't have to say he loved us. We already

knew. He had already shared love—what Csányi calls "this miraculous ability"—with the Root family for years. When he died, my family lost a beloved friend.

Still, it's nice to know that scientists Berns and Csányi would argue that we also lost a being who truly loved us back.

12

Bobby and the "Righteous" Dogs of Egypt

I wished I'd known it at the time, because I would have been even more of a fanboy than I already was. I had just started the second semester in my PhD program and had stumbled upon the thought of Emmanuel Levinas, to whom I've already introduced you. What I didn't know at the time, however, was that Levinas had written a beautiful three-page essay on dogs.

What makes this little essay so powerful is that Levinas does something very different from in his more proper philosophical texts. He goes biographical, telling about how during World War II, he was captured by German soldiers and thrown into a prisoner-of-war camp, where torture and forced labor were the norm. Yet he survived. After the war, he wrote dense treatises on the ethics and infinite mystery of encountering the face of another person. Levinas wrote with the urgency of an

Old Testament prophet, seeking God's goodness in a world oppressed by evil, never allowing God off the hook for the blood-soaked soil he had witnessed in the camps.

In this brief essay, however, titled "The Name of a Dog, or Natural Right," Levinas turns to dogs. The setting for his reflections is the prison camps that so systematically stripped him and others of their humanity. Not only did the guards treat prisoners as disposable objects, but the townspeople did, too. He tells of being marched back and forth from prison to work site, passing the townspeople at gunpoint. An ideology of hate had turned his captors into monsters, blind to the humanity of the prisoners, blind to their pain, intent only on forcing them to complete their backbreaking labor.

One day, however, on their march back to prison, a dog ran out from the woods and bounded up to the prisoners. His tail wagging happily, the dog jumped up to lick their faces, bringing into their gray world a blur of energy, color, and affection. Though the dog had never seen the men before, it seemed to recognize them, Levinas said. Even better, the dog seemed convinced that these broken, forlorn captives were amazing people.

For two weeks, after their long hours of dehumanization, the same dog appeared in their midst with his happy greetings. To the prisoners, Levinas recalled, the experience felt like drops of rain on a dry, dusty land. "For him," Levinas wrote of the dog, "there was no doubt that we were men."[1] The dog's joy was a kind of defibrillator to their souls, reawakening them to their true worth.

In his essay, Levinas points out that the dog had no such reaction to a tree, and a very different response to a squirrel. Yet when it saw a human being, the dog knew just what to do. It celebrated what the guards and towns-people refused to acknowledge: the lowly prisoners were beautiful human beings. The men responded, Levinas recalled, by giving their visitor a name: Bobby. Reflecting on the events later, Levinas saw the naming as bearing witness to the shared bond between the dog and the prisoners.

After two weeks, Bobby disappeared, never to return, but Levinas said the dog's sacred work and witness remained, and the prisoners spoke of it often. They knew they were persons of worth and beauty now, and no misery could take that truth away.

Righteous Beasts

This experience led Levinas to conclude that not only do dogs connect with us at a spiritual level, they do so in such a way that we can call them righteous—meaning that a dog is, or can be, in right relationship with God based on its actions. I found that statement amazing. After all, the term *righteous* comes freighted with significance in the Jewish tradition, and Levinas wouldn't have used it lightly.

No surprise, then, that to explain what he means by calling Bobby "righteous," Levinas shifts from philosopher to rabbi, creating something like a midrash (rabbinical commentaries of ancient Judaism) to help reveal a hidden truth. He explores his experience in the Nazi camp in light of a biblical text, Exodus 11, which describes the last of the ten plagues that God visited on the Israelites' captors in order to set them free.

When this last plague strikes the land, the text says, "among the Israelites not a dog will bark at any person or animal. Then you will know that the Lord makes a distinction between Egypt and Israel."[2] These dogs participated in the liberation of the Israelites, Levinas says, by standing witness in silence rather than barking to warn their masters that *slaves* were trying to escape. Scholar Laura Hobgood-Oster agrees; by their silence, she says, the dogs were "proving that they are acting as agents of the divine."[3]

That, Levinas explains, is why they were righteous dogs.

Taking us back to a Nazi prison camp, Levinas explains that Bobby, a stray from the woods, along with those dogs of ancient times, was bearing witness that those oppressed are indeed human. Throughout human history, righteous dogs have participated in God's act of setting people free to be fully human. Bobby couldn't break the men out of prison, but right under the guards' noses, he sprang them loose from lies that threatened worse.

You might have a righteous dog stretched out beside your chair or bed right now, as I did at my feet the first time I read Levinas. In our day, a dog can speak loudly against a culture that defines our worth and value by what we produce and consume, and that differentiates between people as more or less valuable based on the color of their skin, or on their education or income. Dogs don't. To a dog, there is no difference between a person gifted enough to create a Fortune 500 company and someone whose hair is gone from chemotherapy or who is confined to a wheelchair. Human beings, these righteous dogs proclaim, are beautiful. Period.

Interestingly, Levinas believes that Jewish law rewarded the righteous dogs of Egypt for their gift. Exodus 22:31 reads, "You are to be my holy people. So do not eat the meat of an animal torn by wild beasts; throw it to the dogs." Where you or I might read "throw it to the dogs" as a sign of disrespect or devaluing of dogs, Levinas proposes a different interpretation. He views the meat as a blessing and reward given to the dogs for their sacred partnership with God's purposes.

Do dogs all merit being called "righteous"? I don't think so. We read of dogs attacking small children. We see news clips of security dogs foaming at the mouth and tearing the flesh of prisoners or of students marching against oppression and injustice. Yet it seems to me these dogs are behaving the way they've been trained. They have been bent against their own nature for other purposes by broken human beings. Time and time again, though, dogs remind us that we are beautiful simply because we are, and this is a righteous gift.

Whether you agree with Levinas's commentary or

not, I believe this much is clear: a good dog echoes the acceptance and pleasure of God. As a theologian, I find this undeniable. Also, as a father, citizen, and dog lover in our violent and hate-filled era, I find it inspiring. Like never before, we need these descendants of the righteous dogs of Egypt to nudge us with their wet noses, reminding us that we, our children, and yes, even our enemies share an infinite, and infinitely *valuable*, spirit.

13

A Story from the
Cave of Dreams

One night, over pizza, around the dining room table, I told Owen and Maisy about Levinas and Bobby. I wondered aloud whether Kirby might be a descendant of the righteous dogs of Egypt. To two kids growing up in church, hearing that dogs were both present and helpful during the Exodus turned out to be exciting news. It didn't take them long to reach a verdict: though not all dogs might be descendants of the righteous dogs, Kirby surely was.

Then Maisy asked, "What is a descendant?"

"Well, you are a descendant of your ancestors," I explained. "When you were born, you came from your parents and from their relatives who came before them. They make you, in part, who you are." For example, I explained, she was a descendant of her great-grandma (aka G.G.).

With this information in hand, Maisy pressed on: "Then who are the ancestors of the righteous dogs of Egypt?"

"Wolves," I said matter-of-factly. My five-year-old, though, sensed that I had jumped the rails of my "Maisy to Mom and Dad to Great-Grandma" example.

"But, Dad," she protested, "how did wolves become dogs?" She wasn't sure she believed me, and I wasn't sure how to explain. I would need to dig further.

Now that I had come to see the bond with dogs as potentially profound, even spiritual, I, too, needed to explore the question of origins: How did a wolf howling at the moon transform into a black Lab fixated on tennis balls? Where did the human relationship with canines start, and how?

By then, it had been weeks since Kirby's passing—just long enough for the tears over him now to come only at bedtime. Right around then, the vet's office called to inform us that the box containing Kirby's ashes was ready for us to pick up. Along with the box of ashes, the vet gave us a plaster mold of Kirby's paw print.

I picked up the print and inspected it like a prospector in the California gold rush peering at a chunk of ore. I traced with my finger each part of his paw print, every unique bump and twist and nail print that made up his pad. I remembered how Kirby would place that paw on me over and over again every afternoon as he begged to go outside and play ball. Now all that remained were his ashes in a box, a few strands of hair, and a paw print in clay.

Yet Kirby's print reminded me of something else: the paw prints I had just read about, pressed into the clay on the floor of Chauvet Cave in southern Europe. The enormous Chauvet–Pont d'Arc Cave, first discovered in 1994, reveals evidence of human habitation and, more significantly, human consciousness from 26,000 to 32,000 years ago.

Deep underground in this cave, paintings on the walls depict a dreamlike menagerie of beasts, most now extinct (cave bear, giant horse, bull, woolly rhinoceros, mammoth), along with the outlines of human handprints. Unlike most cave art, the images do not seem to connect directly to hunting rituals, but to something higher. For the first time in their existence, hominids seem to have created spaces for the imagination, perhaps for worship and prayer.

Scholars say the paintings strongly suggest that *Homo sapiens* could now think about thinking, a remarkable accomplishment in itself. Yet what captured my imagination most about the discoveries at Chauvet was smaller, more personal. Pressed into the fossilized mud in the cave are the footprints of a boy. He is about the same age as Owen. He seems to be on a stroll through the cave, but he is not alone. There, beside him in the clay, are the paw prints of a wolf-dog.

Right at the time and place where humans are

revealed as spiritual beings, a wolf-dog appears—and not at a distance, but accompanying a boy, perhaps guarding him.

The prints tell two stories, but of one journey.

Walking into a Shared Future

The prints in the mud reveal what appears to be a great leap forward at the same time for dogs and humans. Across the globe, archaeologists have found an extraordinary number of dog burial sites, some dating back at least fifteen thousand years. It appears that wherever there was culture, there were dogs. Archaeologists have even found a number of human graves with dogs in them, adorned to communicate clearly that this was no accident or mass grave, but an intentional burial of *this* human and *this* dog, highlighting the special bond between them.

The narrative that first appeared in the ancient clay is still being told in our neighborhoods. In the far corner of our yard in St. Paul lies a mosaic plaque that our children made for Kirby. It marks the place where our dog's ashes rest with his favorite tennis ball and bone. The morning we put Kirby's ashes in the ground, we stood together as a family and prayed. We had no idea we were doing something other *Homo sapiens* have done since the last ice age—it just felt right to give Kirby back to the earth in a reverent manner. Yet, that day in our yard, it turns out we were participating in something as old as Chauvet.

How did such an unusual bond between species come about? That's what I want to explore in this chapter.

Unfortunately, the bones in the cave, the paintings, the intermingled tracks in the clay, suggest only at what the boy and the wolf pup were doing there. What we need, our friend Levinas might propose, is something like a midrash: an informed commentary that can bring hidden truths to light and help us fill in the gaps of our understanding.

For the story of the cave at Chauvet, and begging pardon from those rabbis of old, here's mine.

A Midrash of Wolf-Dog and Child

A long, long time ago—I'll say thirty thousand years ago—a boy stepped out of the brightness of the day and into an opening in the side of a hill. The boy was maybe eleven years old, and he took with him into the dark of that cave a burning torch to light his way and a silent companion to guard his steps. His companion was a wolf-on-its-way-to-a-becoming-a-dog.

The boy[1] with the torch was on his way to becoming something, too. I think that's why the underground rooms full of art he'd heard so much about drew him onward so powerfully, deeper and deeper into the cave. His people had found that the stillness and beauty of the art in these caves mysteriously moved them to connect with a higher Being, a Creator Spirit. We could think of the wolf-dog[2] walking behind him, leaving its own paw prints in the same soft clay, as the precursor of the righteous dogs of Egypt, journeying with that child into the spiritual ritual of his people.[3]

Not that you and I would be surprised by this little scene. From our vantage point eons later, we know how dogs and people have shared in this kind of deep connection. Yet, on this particular day, the boy and his wolf-dog companion were making an evolutionary crossing that would be memorialized in the cave floor. *Homo sapiens* seem to appear from hominids as if for the first time in the flickering light of that cave, and the dog (*Canis familiaris*) also seems to appear for the first time, almost in the act of parting ways with the wolf (*Canis lupus*). We have to ask, what brought the boy and the beast to make their footprints so close together? For the answer, our midrash will tell a story of a shared meal and, before that, a shared hunt.

Sharing in the Spoils

As the brain of the boy's evolutionary ancestors continued to enlarge, protein in meat became increasingly important. The beasts then roaming the tundra and Ice Age forests offered all the meat any hominid or wolf could want. Mastodons, woolly mammoths, rhinoceroses, giant deer, long-horned bison, musk ox, and horses—all these were enormous (megafauna) by comparison to today's mammals.

Yet, looking to megafauna for dinner proved a challenge for the boy and his parents. Even with flint and stone projectiles, most hominids would need to be very close and very lucky to take down such beasts. Likely, most hominids came by the majority of their meat through scavenging from carcasses of animals taken down by larger predators: cave lions, panthers, bears, and cave hyenas.

This quest for food is likely where the boy's kin first came into close and frequent contact with wolves. Though it was not at first apparent, wolf and hominid actually had much in common. They were relatively similar in size. They hunted and scavenged in packs, where members bonded; had empathy for those in the pack; and played, particularly with their young. Both species were willing to fight and die so their young would survive.

Still, whether a carcass was freshly killed or scavenged by one species (wolf or hominid), soon the other would come close looking for parts of the carcass to steal. The boy's great-great-grandfather may have eaten his scavenged meat knowing it was stolen from a wolf chased away at spear point.

Unlike the disgusting regurgitating process of the wolf, Great-grandfather and his people needed to drag the meat back to their children. It is likely that wolves began to follow, first standing in the distance, breathing in the intoxicating odors of Great-grandfather's prehistoric cooking fires (by then, his people had mastered fire).

In the midst of this survival dance, wolves must have

taken on mysterious powers to humans. They showed up at every turn, staring from afar, watching curiously and with evident intelligence—perhaps to plan an attack or a raid, or perhaps just to observe. One day, our midrash tells, the boy's great-grandfather looked up to see a wolf gazing at him in silent contemplation, as if to read his gestures and understand his intentions. At that moment, he was gripped by a sensation that the wolf could read his mind.

I know that feeling. When I first started dating Kara, she lived in a small bungalow in an urban LA neighborhood. The owner of the house had five dogs protecting her property. One of the dogs was named Dingo because he looked just like a wild dingo. Whenever I came over, three of the dogs would bark in delight and one in apprehension. Dingo, however, would sit quietly and stare at me as I walked through the yard. His wolf-y gaze always left me with a deeply eerie feeling.

I think it was that sense of shared presence in the wolf's gaze that finally prompted the boy's great-grandfather to take action. It happened on a day when a wolf sat watching, and maybe howling, closer than usual, and on this day, his great-grandfather did *not* chase it away with his spear. Instead, he turned and threw the wolf a piece of meat.[4]

Why? I think by then Great-grandfather already felt oddly bonded with the wolf. He had grown accustomed to its presence, seen it romping with its pups, and sometimes, when the horizon was empty, he found himself wondering where the wolf had gone. Before long, the meat toss became Great-grandfather's ritual. And the more he did it, the closer the wolf would come. Increasingly, the watching and waiting turned into more shared food and closer proximity.[5] Like pigeons eating sandwich crumbs in Central Park, shared food made the wolves tamer, more accustomed to human nearness, and eventually more dependent on us.

Now imagine a day when a litter's mother was killed. Suddenly, the pups found that food had stopped showing up at the mouth of the den. Soon enough, with an empty stomach, at least one pup waddled its way right into Great-grandfather's camp following the smell of dinner. Naturalist Mark Derr asserts that the volition to domesticate most other animals in the human story came from our desire. The wolves that became dogs are the only animals that came to us, crossing their own boundary of fear in order to reach out in friendship to us. For a generation or more, our midrash says, Great-grandfather's people had been ritually feeding wolves, but the animals would take the meat and run. This time, the pups ate and stayed.

Over generations, the bond between human and canine became normal, and neither side wanted it to end. More significantly, the wolves' new life with humans began changing their minds. They were learning to read human faces, words, and gestures, discovering the benefits of cooperating with our actions and, above all, coming to be infatuated with us.

This relationship was changing humans, too. In a world without cities and villages, where we were still part of the food chain and always in danger of a bear attack or a Neanderthal raid, having partnership with wolf-dogs with extensive capabilities of scent and hearing would have been invaluable. The wolf-dog puppies who grew and stayed became our alarm system. Scientists Raymond and Lorna Coppinger explain how this would've worked:

> Stalking a prey, that is, sneaking up on it, while something is barking at you just doesn't make sense. In most cases, approaching and barking at a predator are enough to divert its attention away from the hunt. Most carnivores can't continue hunting if some dog is yapping at them.[6]

Yet this alarm system did much more than keep us safe. It allowed us to cultivate a particular basic resource that we couldn't think of living without today. This resource is so simple, yet so deeply important, that in only forty-eight hours without it, our minds will seize up like an engine without oil. I'm talking about sleep.

In a dangerous world, uninterrupted sleep would have been hard to come by. To make sure the tribe was not the next meal for something out there in the dark, one or more adults would need to keep watch. Even then, fear of predator or hostile hominid would have stolen sleep on many nights. Partnership with wolf-dogs as sentries would have come as something of a breakthrough, inviting for humans the REM sleep in which we dream and renew our minds.

The spiritual traditions of hunter-gatherers gave particular credence to dreams, as do the Old and New Testaments of the Bible. By watching over them as they slept, protecting them and partnering with them, the wolf-dog allowed our forebears to expand their minds into deeper realms. During waking hours, *Homo sapiens* had the luxury of entering into the deeper contemplative state of daydreaming because the wolf-dog was standing guard.

When Owen was little and caught in a nightmare, he would cry out and thrash around, powerless to wake himself completely from sleep. Even after we woke him up, he remained afraid and confused about what was real and what wasn't. We discovered, though, that when Kirby jumped on Owen's bed, nuzzled his chin, and allowed Owen to reach up and hug him, the spell would be broken. Kirby's presence drew Owen back to safety.

Daydreamers and Night Watchers

As humans and dogs evolved together, I believe the presence of a dog during our waking hours became even more important than security at night.

When Kirby died, I found that being alone felt different. As far back as I can remember, I've always spent long hours alone in my study reading and writing. I like being alone, but even when I was alone, Kirby would often be there lying at my feet. Even though I rarely was consciously aware of it, my mind was always on Kirby—not enough to distract me from my work, but enough to notice the absence when he was gone. When he was there, I felt free to let my mind roam and do my work of searching for new ideas about God. Yet in the weeks following his death, without my dog lying next to me, I struggled to get work done. I wondered if I'd ever be able to write again. I told myself I was being silly, but truthfully, it felt like part of my being was missing.

That's why, in thinking about what happened between that child and wolf-dog, I have a hunch about what the two companions were intent on doing in the cave that day. I think they sought out those walls of art to experience something like wholeness, and maybe even worship. Here is what I mean: The boy came as a participant, not just an observer. On the walls of this shrine, he read and responded to his own story of becoming. I can see the flame flickering, hear the burning pitch popping and fizzing in the great cavern around them. Meanwhile, child and wolf-dog watch silently as the mysterious representations of giant beasts seem to move across their field of vision. The wolf-dog's job was to sit in the boy's presence, watching, too, and bringing its own heightened awareness while the human imagination traveled higher through states of meditation, contemplation, and prayer.

Is this exactly how it happened? We'll never know. Yet we do know that there are two constants about human beings. One, from as far back as we can go, we have always had some form of spirituality: a yearning for God, gods, or transcendence. Two, we've always had dogs. Of course, I'm a theologian—the story of God and humans in search of each other is my passion. A painter or ethologist might tell a different story, or word it differently. Still, given what we know about the ancient humans who etched pigment and charcoal into the walls of those caves, I find hope in the thought that the deep, spiritual connection I felt to Kirby, far from being a delusion, is something we have experienced with dogs ever since our beginnings.

And if that connection stretches back to the furthest reaches of our past, there might be reason to believe it will stretch forward into eternity.

14

The Grace of Dogs

A few months after Kirby's death, Owen called me to his room. Bedtime was still hard for him, and he wanted to talk. Looking at me with tears running down his freckled cheeks, he asked me a question I was surprised he hadn't asked before: "Will I see Kirby again? Will Kirby be in heaven?"

Like most parents in that situation, I knew what I *wanted* to say. "Of course. Absolutely! No doubt." In moments like that, every parent wants to bring comfort. Yet, as these pages have pointed out, there's much more to it than that. After all, if we do have a spiritual connection with our dogs, wouldn't it make sense to expect that they'll meet us in eternity? For that matter, isn't it because of the deep connectedness we feel with our dogs that so many of us wonder whether we'll see them again in some place beyond death?

Still, in that moment at Owen's bedside, I wasn't sure what to say. "Mmm, wouldn't that be wonderful?" I offered lamely, and tried to change the subject. Yet I wanted to give my son more than a hopeful shrug or a sentimental fantasy—and secretly, I was wondering the same thing. Where could I look for an answer?

The story of a brilliant thinker who was humbled by the same question opened a new way of thinking for me.

In 1928, more than a decade before participating in a plot to assassinate Adolf Hitler, the renowned German theologian Dietrich Bonhoeffer was a pastoral intern in Barcelona. He had just finished his PhD (at the age of twenty-one, no less), and was recognized by the prestigious faculty of the University of Berlin as a theological prodigy.

One day, a ten-year-old boy came to see Bonhoeffer. Breaking down and crying, the boy explained that his beloved German shepherd, Mr. Wolf, had just died. The boy sobbed as he told the story, but soon his tears stopped and he asked Bonhoeffer, with deep intensity, "Tell me now, Herr Bonhoeffer, will I see Mr. Wolf again? He is surely in heaven?"

Bonhoeffer explained in a letter to a friend that he was dumbfounded. He didn't know what to say. Never before had one of his astute professors or gifted fellow students made such an inquiry, a question that Bonhoeffer could see meant so much to this grieving boy.

Bonhoeffer sat with the boy, feeling small next to his important question. Clearly Mr. Wolf had meant so much to the boy. The overly confident protégé, who had always been told he had a brilliant answer for every theological question, now sat humbled by the boy's love for his dead dog.

Finally, turning to the boy, Bonhoeffer said, "Well, we know you loved Mr. Wolf, and we know that God loves you. And we know that God loves all the animals. So, yes, yes, I think you will indeed see Mr. Wolf in heaven, for I believe that God loses nothing that God loves."

God loses nothing that God loves.

Bonhoeffer's point was that when persons relate in love, it is an eternal act that transcends biology, chemistry, and history. Only love lasts forever. Spirit returns to its source in a personal God. Because "soul" is based in our personal relatedness, rather than being some kind of isolated substance, any being that participates in this kind of loving relatedness belongs to God, and will return to God.

This is the grace we are offered, in this world and the next. Grace is the invitation to share in the mind and heart of God; it can never be earned, but it comes to us always as a gift. So maybe heaven is a "place" of personal relatedness, where our relationships can never again be interpreted or corrupted by death, fear, or hate. Maybe heaven is where we are free from all that might threaten our sharing in the mind of one another and God. It is where our bodies are free from all that could upend or damage what we yearn for most, to be shared in and share in others.

All who participate in the gift and grace of deep personal relatedness are never lost. God will never let them go; for this deep connection rooted in love does not disappear when a loved one dies. When Jesus on the cross entered death, he built from within it our relationship with God and with one another, a relationship rooted in a love that is eternal and over which death no longer holds power.

These days I tell my son that he *will* see Kirby again. Like all those who have shared in his mind and heart through love, I tell him, Kirby will be resurrected again through the power of the God who is the eternally personal relationship of Father, Son, and Holy Spirit.

This is a theological conviction that I hold by faith. I teach it to my son so that Owen can feel his way into this reality, can begin to have words for the experience of love and grace.

Of course, reality is always putting our convictions to the test. At a deeply emotional level, whether we are people of faith or not, we all struggle with the grief of absence and wonder where our loved ones go when they leave us. If, in a moment of doubt, you were to press me to answer the question "Do you *really* believe that Owen will see Kirby again?" I might answer more tentatively. Still, I know in my soul—like Owen on the floor of the vet's office, like the boy long ago in that Chauvet cave—

that the love of a dog is strong enough to last both in this world and in the next.

"No one knows for sure," I'd tell you. "But I've studied this, and I think so. My answer is yes. The grace of God is echoed by the grace of dogs. And grace is eternal."

15

Khaleesi Arrives

The way to end a story you want people to read, especially one about dogs, is with a happy outcome. Old Yeller may have been shot in the corn crib after getting rabies because he saved his boy, Travis, from a rabid wolf, but the Disney movie ends with one of Old Yeller's pups mischievously stealing his way into Travis's broken heart. And while it's true that there will never be another Kirby, it was our experience with Kirby—and the desire to honor how Kirby had loved, shaped, and enriched us—that got us thinking about getting another dog.

After Kirby's ashes were buried in our backyard, I began to do something I told myself I wouldn't do.

I knew I needed to grieve Kirby. I needed time, I told myself. Plus, I was sure no other dog could replace him. Or maybe it was the fear that I would actually love another dog and forget Kirby. Nevertheless, I found myself online, searching for dog breeders. I told Kara that this time we should get a female yellow Lab. Going opposite in color and gender to Kirby seemed a nice way of keeping our memory of him. The more I searched, the more right it felt.

Months passed, and with each week, I became more certain: I wanted us to get another dog. I wanted Owen and Maisy to bond again with a dog. Kara was afraid I was going too fast—what if we'd just gotten lucky with Kirby? Naturally, I drew on my research on dogs and spirituality to make my case. I told Kara that our kids needed a new puppy because dogs echo a deeply spiritual reality, connecting us to empathy, bonding, and play . . . in other words, pretty much reciting to her what I had just invested months of work to learn, and I had the Amazon bills to prove it. How could she resist a little cognitive ethology?

Turns out she could, and did.

When my intellectual justifications fell short, I went back to the basics: Our kids had never helped raise a puppy; they knew life only with an adult dog. Having a puppy would be enriching to them and would grow them as spiritual human beings!

Again, she looked at me like I was full of it.

A puppy in particular, she reminded me, would drop

a ton of chaos into our lives. My stomach turned when I thought about standing outside in the middle of the night teaching the puppy to pee in the grass and not on the carpet, but in the end, the real reason I wanted a new puppy was that I wanted Owen and Maisy to have another dog to play with. I wanted a dog who would celebrate Owen's and Maisy's presence with excitement, always reflecting to them that they were beautiful and worthy of attention. I wanted a dog who would summon them into shared play, helping them rest in the knowledge that they are wonderfully made. (You can see that I couldn't help but drift out into the theological deep.)

Finally, Kara agreed to think about it.

On a family road trip, I knew we'd pass within a couple of hundred miles of a breeder, so I had taken his number with me. With Kara open to the idea of a puppy, I asked Owen and Maisy what they thought about it. The backseat erupted with instant agreement, so I told them that we'd spend the trip considering it together. This initiated their campaign of relentless pleading and persuasion that became the nonstop soundtrack of our vacation.

Within hours of announcing our discernment process, Owen became the biggest advocate for a new dog, thinking of names, promising to help, and talking about the possibility incessantly. We'd waited some time before calling the breeder, and were almost certain all the puppies would be claimed, so finally, we told the kids that if we called and there was one girl left, we would take her. We would check, but there were no guarantees, and that while we were ready to get a new dog *some*time, this might not be the time. Maybe the decision would've already been made for us.

Kara called Brad the Breeder in Brainerd, Minnesota (cue the Coen brothers), and asked if he had any puppies still available. Of course, it turned out that he had just one unclaimed white female. Like déjà vu back to the day we chose Kirby, Kara found herself blurting, "We'll take her!" When we told Owen, he shouted, "Prayers do come true!"

We now needed a name. The kids had all sorts of stupid ideas, but having just finished a week of putting our kids to bed early so Kara and I could binge-watch season three of HBO's *Game of Thrones,* we thought the perfect name for the pure white female Labrador would be Khaleesi, the white-haired, badass Mother of Dragons princess. To our surprise, and without having any familiarity with the character, the kids agreed. Khaleesi she'd be.

We cut our trip short to drive to Brad the Breeder's house, excited to pick up our puppy. We stopped by PetSmart and bought a pink puppy collar, leash, and a small kennel we could drive her home in and use as her bed for the next few months. That night, we stayed in a hotel, preparing to pick up Khaleesi the next morning. I got up early to put together the kennel out in the back of the car, trying my damnedest to assemble the dumb thing in the hotel parking lot while projecting an "I've got this" confidence whenever anyone walked by.

As I was wrestling with it, a couple talking in the deepest southern accents I had ever heard approached the car next to mine. Seeing me fighting the kennel, they asked each other with enough volume to reach the entire parking lot, "What's *he* doin'? Oh, I see! He's puttin' together a *critter bucket*! But I don't see no *critter* for that *bucket*! *Where's* the *critter?*"

I finally got the bucket together, and we were ready to pick up our critter.

An hour later, we pulled off the country road and into a gravel driveway. Maisy was the first into the garage of Brad the Breeder. Before I could get in the door, she was inside the pen, being mobbed by a pile of the cutest things I'd ever seen, a mess of tiny pink tongues, black noses, and wiggling white-yellow bodies. Maisy shrieked with joy as the slew of roly-poly, velvety puppies crawled over her and kissed her, going, of course, right for her face.

Brad picked up the one with a green collar and handed her to Owen, saying, "Here she is; here's your new dog." Owen's face lit up so bright that it looked as if his hair could stand up straight. And the white, eight-pound Khaleesi sniffed his chest, stretching her little body to reach for his chin, sniffing at his face with her wet nose.

Seeing that puppy in his arms, Kara and I both fought back tears. A child and a dog. It felt right, oddly holy, and a deep privilege to witness.

On the Journey Again

And three years later, here we are on this journey again.
Khaleesi has grown to be a fabulous dog. Her preferred
way of greeting us is to press her face against ours with
her eyes half closed, breathing deeply and contentedly.
She decided early on to sleep every night with Maisy.
Each evening, she lies in the middle of Maisy's bed with
her paw thrown over Maisy's legs and her head resting
beside her. Every morning, she bounds into our room
and burrows between Kara and me to cuddle and say
good morning, to welcome us in joy to a new day filled
with possibilities. Every afternoon, I see her face in the
window watching for my return, and at the door, her
happy eyes and frantically waving tail greet me excitedly.

I drop to my knees and bury my head in her neck,
feeling her happiness against me. It is grace.

ACKNOWLEDGMENTS

This book has been no small task in writing. It has consumed large parts of almost three years of my life. Before starting it, if someone had asked what kind of project I'd work on for three years, I would have assumed some five-hundred-page academic tome with two thousand footnotes—not a short book about dogs for a general readership, for goodness' sake! Yet, so it has been, and so I've learned that writing (true writing that seeks to reach a reader more than prove a point) is a bitch! A blessed bitch, no doubt, but a monolithic task nevertheless.

Speaking of bitches, at first blush, writing on dogs might seem odd for me. I haven't cut my teeth writing on Saint Francis or ethical and religious dimensions of ecology or animals. Yet, in another sense, this direction fits me perfectly. I'm a practical theologian, seeking to explore the theological and spiritual depths of our lived experience. For many of us, there is no more concrete lived experience than that which we have with our dogs. I've also spent a majority of my life researching and teaching the spirituality of children and adolescence. While this book on dogs doesn't seem directly connected to that, either, the inspiration for delving into this lived experience of the love and loss of our dogs was born out of the experience of my own children and the wonder of what it is that dogs do for young people.

So it is first to them, Owen and Maisy, that I must offer my thanks. Too often I've shared stories about them in writing and presentations without their permission. Yet this project was different. They cheered me on, offering their most painful experience of loss for others to read. Writing about dogs actually made me a minor celebrity to Maisy and her friends. Often Maisy would say to classmates, "My dad is writing a book about dogs," and the other kids would *ooh* and *ahhh*. That's a rare experience for a theologian. She saw me as an expert in something that matters deeply to her. The downside was that she asked me repeated questions, such as "Can dogs see green? Can they really smell fear? What is the largest dog ever?"—many of which I did not have clear answers for, making me feel more like an impostor than an expert.

This book took three years to write because I rewrote it at *least* three times (with major revisions in between). Luckily for me, I had an honest coach in so doing. Kathy Helmers has been not only a literary agent; she is also a confidante and truth-teller. More than once (more like half a dozen times), Kathy told me the manuscript wasn't good enough and to push harder. Yet this criticism never came without direction, inspiration, and assistance. This project would never have come to be without her patience and willingness to struggle along with me. After much sweat and blood, Kathy convinced David Kopp to take a shot on it at Convergent. It has been an amazing pleasure to work with David. His kindness and direction have been inspirational. I'm also deeply thankful to Derek Reed, who provided important insights on the project and helped me to make it so much stronger.

A few friends also deserve direct, in-print thanks. Tony Jones encouraged me in this project and connected me with Kathy at the beginning. Carla Barnhill lent her amazing editing skills to an earlier version. And David Lose was the first to hear about my hope to write this, encouraging me, during a run together, to go for it, and assuring me that it wasn't crazy.

Finally, there is one person who deserves the largest thanks, and that is my wife (and partner in all things), Kara Root. This is no proverbial "thanks" to the spouse, your support and blah, blah, blah helped me through . . . Rather, Kara's help went much, much deeper. At least for two versions of the rewrites, she sat next to me and worked out the edits, helping me write things more clearly, directing me. Truth be told, Kara is a way, way more talented writer than I am, and her genius can be found on all these pages. So, with not one shred of hyperbole, in all honesty, I say that this manuscript and our shared experience of losing Kirby would not have been possible without her. She has been the great blessing of my life.

APPENDIX

"Do Dogs Go to Heaven?"

and Other Questions

As I mention in the early chapters, I balked more than once at telling people, even friends, about this project. What business could a relatively sober-minded theology professor have with dogs? For a time, I considered writing under a pseudonym. I could be like Søren Kierkegaard, the melancholy Danish philosopher who regularly angered Denmark's elites by pointing out their hypocrisy while writing under names such as Judge William, Johannes Climacus, and Anti-Climacus. I'd heard that the best pseudonyms are your middle name and the street you grew up on. Mine would be James Camelot, then, but that just didn't seem credible, though I did like the thought of doing interviews in a bad British accent.

Reluctantly, I began trying out my thesis with some friends. These conversations always started the way I feared. They'd ask, "So what are you working on?"

"I'm about finished writing a book on . . . [cough] . . . dogs," I'd say.

"What? Dogs? Why?" The responses were always the same.

Invariably, though, after ten minutes and a few stories about our spiritual connection with dogs, my friends would be hooked. "Wow," they'd say, "that makes so much sense. It hurt so much to lose our dog." Or, "Amazing! I guess I can totally see that. I loved my dog so much growing up." And often they'd start sharing their own stories.

In case you don't have a theologian in your family circle, here are the questions I heard most frequently, along with my answers—all in one place.

Do you think dogs have a soul?

My answer is *Yes, of course!* And also *Well, that'd be weird.* There have been countless volumes, just within the Christian tradition, written on the concept of a *soul*. It's actually been quite a tricky conversation. The concept of the soul assumes that there is something in us that is nonmaterial, something that transcends or resists reduction. We think of our soul as the essential part of us that makes us *us*. It's also the part that makes us morally responsible.

Too often the idea is that the soul is an eternal ghost inside you that leaves the body when it dies—scenes from old cartoons might come racing to your mind. Yet the biblical traditions of both Judaism and Christianity have always been a little uneasy with this "ghost in a body" view of the soul; it's actually an idea that comes more from Greek philosophy than the Bible. Also, lately, science, too, has made it hard to see the soul as some hidden ghost inside us. The more we learn about neuroscience, the more we recognize that our moral reasoning and personalities are actually dependent on our brain (not on a ghost). If your brain is damaged in an accident, it can radically change your personality.

It seems more faithful to the theology of the Bible (and even new brain science), then, to see the soul not as a hidden ghost but as a capacity to relate. Human beings are soulful because we have the ability to share deeply in the life of another person, feeling what they feel, standing for them, and even sacrificing so that they might live. We have unique personalities and moral responsibilities because of our deep forms of relatedness. We embrace the soul of our children because our connection is so deep. Soul is the palpable spirit of relationship. It is nonmaterial (there is no material location for love), but it is accomplished only through our material bodies.

So, do dogs have souls? If we see the soul as being based in a form of relatedness, then we'd have to say yes. Dogs are uniquely shaped to relate to us through our embodied soulful center, our face. The dog may lack many of the cognitive abilities of humans, but the dog, like no other animal, relates to us in a deeply personal way. This ability makes dogs deeply soulful beasts—not because they have a hidden, eternal ghost within them, but because, like infants, they give themselves to humans in a way that can best be called soulful.

Do dogs sin, then, or are they innocent?
Sin, too, is a tricky concept in my theological tradition. Counter to popular belief, sin isn't really *doing bad things*. It's deeper than that. Sin, like the soul, is bound by relationship. We sin when we violate the relationship between us and God, between us and others. To lie, cheat, or steal is sin

because it violates the relationship between you and your neighbor. We sin against God when we deny that we are our brother's keeper, when we ignore the interrelatedness between us. We sin when we refuse the Word of God that calls us to care for the world. It impacts our soul (not the ghost within us) because it threatens and/or destroys the relationships between us. Sin is the blockage that keeps us from engaging in that deeper form of relationship.

So, do dogs sin? Well, we've seen that they have a deep form of personal relatedness. So, I suppose it is possible that a dog could violate this relationship. Anyone who's been bitten by a dog they've loved would say they felt betrayed. Yet, it would also be unfair to throw too much responsibility on a dog. Dogs are these unique creatures that relate to us in a very personal way, sensing and even yearning for the soulfulness of our personal relationships. But dogs can go only so far in this relationship. It is a spiritual and soulful relationship because it is face-to-face and filled with empathy, bonding, and play. But dogs also lack the ability to speak, and the mental capacities for deep forms of reasoning. So, yes, dogs are *kind of* able to sin, but not quite like human beings.

What is beautiful, though, is that even when doing something wrong, dogs ask for and accept our acts of forgiveness, as many of us have experienced. They droop their heads and tails, look up at us through sad eyes with their ears pressed back—all gestures that seem to communicate both some awareness of their transgression and a longing for the relationship to be restored. I take this as a spiritual act, and an indication that they do have some kind of moral compass, an ability to read right or wrong, good or evil.

Maybe Bill Murray said it best when he tweeted: "I'm suspicious of people who don't like dogs, but I trust a dog when it doesn't like a person."

Will dogs live eternally in the afterlife?

When a family pet passes, most moms and dads hear some form of the "dogs in heaven" question. The humorist Will Rogers may have offered my favorite reply. "If there are no dogs in Heaven," he once said, "then when I die, I want to go where they went." Without benefit of philosophy, science, or sacred text, Rogers went straight to the heart of the matter: the irrepressible child in each of us is intuitively ready to use a beloved dog's eternal destination as the measure of God, goodness, and, yes, our hope of heaven.

I share my replies to Owen's question in chapter 14, but to give a more complete answer, I want to dig deeper into the Christian tradition.

As understood by most in the mainstream of the faith, eternal life is not necessarily heaven as the perfect place, like the Hilton Hawaiian without the crazy meal prices. Rather, to experience eternal life is to be within the relationship of the Father, Son, and Holy Spirit. It is to be so deeply in the love of relationship that even death cannot interrupt it. God's love is so deep that it enters into death, breaking death, so that even death leads us back into a relationship with God. In chapter 14, we learn from Dietrich Bonhoeffer that God doesn't lose anything God loves because God binds what God loves in relationship. And a relationship with God is eternal. So can we claim that our dogs, too, are in this relationship?

It is clear without doubt that you and your children love

your dog and that your dog, in turn, loves you. This unity of love has an eternal dynamic. Jesus promises that in his father's house (in eternity) there are many rooms (John 14:2). And in this house, all tears are wiped away, because all relationships of love and deep face-to-face sharing are restored. Heaven is such a place. We've seen throughout this book that our relationships with our dogs stretch to this deep level. God is love; all love comes from and returns to God. In the same way that we will see again our grandmother, our brother—those with deep inner relationships to us that are so imprinted on our being that they can't be gone forever—in the same way, it is logical, and not simply sentimental, to assume that we will see our dogs again.

So, then, we can claim that dogs will be swept into the eternal, not because they're just so cute and good-natured, but because in their face-to-face love, we experience a true manifestation of the relationship of God as Father, Son, and Holy Spirit. They are walking, bouncing, wagging witnesses to the eternity we sense only fleetingly in our earthly day-to-day.

At least that's how I think of it. Will I see and know Kirby again in heaven? I think so. I can't be sure, because none of us is sure how eternal life works. Yet I do believe that just as I'm not lost to nothingness when I die, so, too, Kirby will live on with me in the mind of God, who is the relationship of Father, Son, and Holy Spirit. And I can live with that.

We've seen in this book that dogs should never be understood as objects. When objects are gone, they're lost forever. I am under no illusion that I'll be reunited with all my old MacBooks in heaven. Yet relationships are different, and dogs, as we've seen, are more than furry objects. The love they give us is an echo of eternity, and all that loves in a personal way is never lost, for it has its origins in the being of God.

So . . . what about cats?
To hell with them! (Just kidding. Maybe.)

But what about other animals? Do they too have spiritual connections to us?
I think it is without doubt that human beings connect to other animals, and this connection of human beings to other animals is a deeply spiritual one. Loving a cow, pig, or bird can open us up to transcendence, reminding us that we are spiritual. I would never minimize that. I'm no cat lover, but I might even say that caring for a cat could do the same. Yet what I've tried to point to in this book is the uniqueness of dogs. What's unique about dogs is that our connection with them is a two-way street. It might be that another animal could do a very doglike thing and connect to us in this face-to-face way, reading gestures and showing empathy. Yet what's amazing, and what I've tried to show in this book, is how dogs do this more reliably than any other animal.

Other animals also have a place in the economy of God. There are ways that God uses them, and they are deeply loved and enjoyable to God. (That's part of the message of the book of Job.) Yet there is just something *unbelievably* unique about dogs in the fact that they so regularly do for us what we so desperately need: remind us that we are worthy of connection, and promise us in their love that there is hope, compassion, and mercy—in this world and in the next.

NOTES

3. BOUND BY SPIRITUAL TIES

1. Emphasis added. Quoted in Darcy Morey, *Dogs: Domestication and the Development of a Social Bond* (London: Cambridge University Press, 2010), p. 2.

2. Lorenz doesn't come to us without problems. In the late 1930s he joined the Nazi Party, accepting a university chair. Much like Martin Heidegger, often considered one of the greatest philosophical minds of the twentieth century, Lorenz's legacy is marred by these political associations. Yet, like Heidegger, his academic work has had an important impact.

4. MINDLESS, FURRY MACHINES

1. Berns gives some texture and pushes in directions I'll be heading here: "Instrumental learning forms the basis of every dog-training method ever published. Teaching the 'sit' command is based on instrumental learning. Here, the stimulus is either a hand signal or a spoken

word, and the desired behavior is the act of sitting. When the dog sits and he is immediately rewarded, he makes an association between the act and the reward. In instrumental learning, the link between stimulus ('sit') and act (sitting) is called the stimulus-response (S-R) relationship. Instrumental learning is also called operant conditioning because the animal learns to *operate* on, or affect, the environment." Gregory Berns, *How Dogs Love Us: A Neuroscientist and His Adopted Dog Decode the Canine Brain* (Boston: New Harvest, 2013), p. 35.

2. John Homans, in his popular dog book *What's a Dog For?*, nicely summarizes this "LEGO theory": "Comparative psychologists see the mind as a toolbox with a whole range of particularized modules—some of them fairly complex and specialized—that fit together in specific ways. By contrast, behaviorists like Wynne see it more as a LEGO set. Their aim is to break complex behaviors down as if they were LEGO constructions, to see how they're linked, finally getting down to a pile of neat blocks. 'I believe that the essence of science is to analyze something that looks complicated into its simpler parts,' he told me. 'Take a behavior like an animal following a human point. I think it can be disassembled so that you see that it's basically a form of associative conditioning.'" From John Homans, *What's a Dog For? The Surprising History, Science, Philosophy, and Politics of Man's Best Friend* (New York: Penguin Press, 2012), pp. 90–91.

5. IT HAPPENS ONLY FACE-TO-FACE

1. Ibid., p. 110.

2. Michael Tomasello discusses further the shortcomings of apes and pointing in *A Natural History of Human*

Thinking (Cambridge, MA: Harvard University Press, 2014), see pp. 52ff.

3. "As insightful as it is about dogs, Hare's tameness finding is most remarkable as a window onto the talents (and lack thereof) of chimpanzees—and, by extension, of humans. It shows the dog to be, in this respect, more humanlike than the chimp. For an explanation, Hare looked to experiments where chimps have to cooperate. He describes a classic cooperation experiment in which a pair of chimps have to pull on opposite ends of a rope in order to receive a reward. But due to their competitiveness and aggression, many of these animal Einsteins fail to solve the problem. In the chimp universe, you can't get there from here. An alpha chimp will simply take a lesser chimp's food, and the weaker one has no incentive to cooperate. No one gets a banana. The only chimp pairs that successfully solve the problem are ones that can eat together without warring at the same food dish—the tamest sort of chimps." From Homans, *What's a Dog For?*, p. 80.

4. Ibid., p. 48.

5. Hare says, "These studies suggest that dogs interpret your gesture depending on what you are paying attention to. In short, Mike and I concluded that dogs have communicative skills that are amazingly similar to those of infants." Brian Hare and Vanessa Woods, *The Genius of Dogs* (New York: Dutton, 2013), p. 53.

6. Vilmos Csányi, commenting on his own study with wolves, writes, "But the tame wolf is not a dog. It will pay no attention to human speech and is not interested in what people are talking about, and when it hears its name, it will not run to its master." From Vilmos Csányi, *If Dogs Could Talk: Exploring the Canine Mind* (New York: North Point Press, 2005), p. 22.

7. "Subsequently, researchers have raised wolves for the sole purpose of comparing their social skills with those of dogs. The researchers gave the wolves even more exposure to humans than the wolves I tested, but their results were similar to ours. At four months of age, heavily socialized wolves could not use a caretaker's gesture to help them find food, even though the caretaker had raised them from puppies. When the wolves were tested as adults, they needed explicit training to match the spontaneous performance of dog puppies." From Hare and Woods, *The Genius of Dogs,* p. 59.

8. Stanley Coren discusses and gives commentary on this study: "When faced with a manipulation task that they can't solve, dogs will stop, look at the face of the person with them, and try to discover clues as to what to do from the person's actions. In comparison, wolves, even those that had been tamed and were living with humans, do not look at the faces of people for clues as to what to do. Dogs can thus extract more information from human social sources around them simply because they are specifically looking for it." Stanley Coren, *How Dogs Think: What the World Looks Like to Them* (New York: Free Press, 2008), p. 237.

9. Alexandra Horowitz, *Inside of a Dog: What Dogs See, Smell, and Know* (New York: Scribner, 2009), p. 46.

10. Bradshaw says, "Although pointing has become the scientists' favorite experimental tool, it is by no means the only activity to which dogs are particularly attentive. They also follow gestures such as nodding and hand movements much more attentively than most other animals do. In addition, dogs seem fascinated by people's eyes and faces: they will follow the direction of their owner's gaze almost as reliably as they will follow point-

ing." John Bradshaw, *Dog Sense: How the New Science of Dog Behavior Can Make You a Better Friend to Your Pet* (New York: Basic Books, 2011), p. 198.

11. Horowitz, *Inside of a Dog*, p. 149.

12. Csányi concurs: "Dogs are excellent human ethologists, too; they continually observe us, and are much helped in this by the fact that they strongly bond with humans. They also have an excellent understanding of human body language." Ibid., p. 132.

13. Csányi, *If Dogs Could Talk*, p. 54.

14. Quoted in Homans, *What's a Dog For?*, p. 75.

6. WINDOW TO THE SOUL

1. From Luther's *Table Talks*, found in Laura Hobgood-Oster, *Holy Dogs and Asses: Animals in the Christian Tradition* (Urban, IL: University of Illinois Press, 2008), pp. 104–5.

2. Remember, Tomasello was so interested in gestures such as pointing because they appeared to be fundamentally hardwired into human beings. Our villain B. F. Skinner and his theory of behaviorism believed that animals (including human beings) were born as blank slates, essentially, as TiVos straight out of the box, with no programmed connections. Yet the cognitive turn that ended Skinner and behaviorism's reign showed that our minds were not just blank slates awaiting programming. Language itself was too complicated (and quickly learned) to be solely the result of behavioral conditioning. Rather, our minds came with already formed modes. Minds, whether chimp, dog, or human, were shaped and ready for certain tasks and operations.

And one of the most fundamental tasks for human beings is the ability to produce and respond to gestures.

Infants, at an amazingly early age, are able to respond to gestures such as pointing, and very quickly begin pointing out things themselves. The human being, compared to other animals, is pathetically dependent for a *long* time in his or her infancy and childhood. The infant human being can't get places on his or her own—or even stand—for years! This is a remarkably long time compared to most other animals.

Still, while the infant is pathetically immobile, very soon after her birth she is uniquely able to read gestures, and to gesture herself for what she needs. This is her great skill to survive! The infant can't run and grab the milk she needs, but she can read gestures and make her own gestures readable enough actually to enter the mind of her caregiver, calling the mother to her person, to meet her basic needs by taking her needs into the mind of her mother.

The infant's ability to enter the mind of her mother is not done through horror movie–like possession, but through the actions of communion; the infant reads and gives gestures as a way to bind her being to the being of her caregiver—and as caregivers will report, this binding is spiritual. Through the communion of shared gestures, we are given a bond that grows through our face-to-face expressions. Hare says it this way: "Almost simultaneously, infants begin to understand what adults are trying to communicate when they point. Infants also begin pointing out things to other people. Whether infants watch you point to a bird or point to their favorite toy, they are beginning to build core communication skills. By paying attention to the reactions and gestures of other people, as well as to what other people are paying attention to, infants are beginning to read other people's intentions" (Hare and Woods, *The Genius of Dogs*, p. 37). This reading of intentions is the catalyst of spiritual ties. The human child is cared for because of a deep spiritual bond that has linked infant to caregiver, and this linking has its beginning in the innate mindful ability to respond to gestures. Responding to gestures is the first profound building block of spiritual ties. So it is significant that dogs are innately able to respond to gestures at all, actually to enter our minds in some kind of communion.

3. "By the end of the fifth week almost all babies are engaging in visual smiling and their smiles become sustained for increasing lengths of time. They are accompanied moreover by babbling, waving of arms, and kicking. Henceforward a mother experiences her baby in a new way." John Bowlby, *Attachment* (New York: Basic Books, 1982), p. 284.

4. "From the time that smiling to visual stimuli is first established, the most effective visual stimulus is a human face in movement; and a face is made even more effective when it is well lit and approaching the infant, and still more so when it is accompanied by touch and voice. In other words, a baby smiles most and best when he sees a moving figure who looks at him, approaches him, talks to him, and pats him." Ibid.

5. I concur with Kirkpatrick's words here. In many ways, I'm offering a theology that believes just this, that the relational connection between mother and child is a deep analog to God's own relational longing for us. Kirkpatrick says, "Psychologically, I suggest, a worshiper's love for God is more akin to a child's love for her mother or father than to an adult's love for a romantic partner or spouse." Lee A. Kirkpatrick, *Attachment, Evolution,*

and the Psychology of Religion (New York: Guilford Press, 2005), p. 77. While I concur with Kirkpatrick, I'm trying to avoid his reductionism in so doing.

7. THE SURPRISING POWER OF CANINE COMPASSION

1. Douglas J. Hall, *Imaging God: Dominion as Stewardship* (Grand Rapids, MI: Eerdmans, 1986).

2. Homans explains that, "[f]or Hare, the notion of the cognitive adaptation, the new faculty, was now out the window. It seemed that tameness itself enabled the foxes to use an existing cognitive skill to interact with the humans they had become so fond of. This new, simpler formulation had many virtues. It showed how complex behaviors could arise quickly—within a generation or two—and from relatively simple mutations. It also tilled in the picture of neoteny as a process that could change the course of a species, as it had with the wolf and the dog—and may have done with humans." Homans, *What's a Dog For?*, p. 79.

3. Hare adds to this: "Perhaps barking is another by-product of domestication. Unlike dogs, wolves rarely bark. Barks make up as little as 3 percent of wolf vocalizations. Meanwhile, the experimental foxes in Russia bark when they see people, while the control foxes do not. Frequent barking when aroused is probably another consequence of selecting against aggression." Ibid., p. 132.

4. Darcy Morey describes further: "Notably, and again surely not surprisingly to many dog enthusiasts, recent test data indicate that people are often able to classify (recorded) dog barks in terms of their emotional content. And they can do so without knowing the context

in which the barks were produced, and with no previous experience with a given dog breed or in owning a dog. That is, they can identify emotional information by sound alone. That factor surely fosters enhanced communication between people and dogs, and serves, in general, to reinforce their social bond. In fact, dogs in general have a remarkable capacity to communicate with people: 'In so many respects, modern dogs seem to be better adapted to communication with humans than with other dogs.' Accordingly, it seems that dogs belong with people . . . Dogs simply do not thrive when deprived of regular human care and interaction." Morey, *Dogs*, p. 199.

5. Bradshaw says, "In this sense, the capacity for love that makes dogs such rewarding companions has a flip side: They find it difficult to cope without us. Since we humans have programmed this vulnerability, it's our responsibility to ensure that our dogs do not suffer as a result." Bradshaw, *Dog Sense*, p. 171.

6. Hare and Woods, *The Genius of Dogs*, p. 87.

7. One of the major impacts of selection for kindness was that it allowed for shared intentionality, and shared intentionality would lead to what Tomasello calls the *ratchet effect*. You need kindness and shared intentionality to get to this. Tomasello says that apes don't have the ratchet effect, which is why they don't have direct spiritual experiences, because spirit is bound in personhood (soul). This quote, however simply, articulates how cultural and even technological advances may have happened: "Teaching and conformity then led to cumulative cultural evolution characterized by the 'ratchet effect' in which modifications of a cultural practice stayed in the population rather faithfully until

some individual invented some new and improved technique, which was then taught and conformed to until some still newer innovation ratcheted things up again. Tomasello argues that great ape societies do not display the ratchet effect or cumulative cultural evolution because their social learning is fundamentally exploitative and not cooperatively structured in the human way via teaching and conformity, which constitute the ratchet that prevents individuals from slipping backward." Tomasello, *A Natural History of Human Thinking*, p. 83.

8. John T. Cacioppo and William Patrick, *Loneliness: Human Nature and the Need for Social Connection* (New York: W. W. Norton and Company, 2008), p. 165.

9. See Elissa Gootman, "The City as Chew Toy," *New York Times*, December 16, 2011, nytimes.com/2011/12/18/realestate/dogs-living-in-new-york-city-city-as-chew-toy.html?pagewanted=all&_r=0.

10. Lieberman offers a nice definition of empathy: "The word *empathy* was introduced into the English language just over a century ago as a translation of the German word *einfühlung*, which means 'feeling into.' *Einfühlung* was used in nineteenth-century aesthetic philosophy to describe our capacity to mentally get inside works of art and even nature itself, to have something like a first-person experience from the object's perspective. *Empathy* still means something like 'feeling into,' but it almost always refers to our connecting with another person's experience, rather than 'getting inside' an object." Matthew D. Lieberman, *Social: Why Our Brains Are Wired to Connect* (New York: Crown, 2013), p. 152.

11. For the close reader there may be some fear that this kind of empathy or soulfulness leads to forms of enmeshment. Here J. Wentzel van Huyssteen and Erik P.

Wiebe discuss Paul Ricoeur's position, showing how empathy actually can lead to differentiation. "Strikingly convergent with Ricoeur, then, it is through the transcendence of empathy that one gains the ability to separate self from other and to see the other as fully other in relation to the self. Through the transcendence of imagination, one receives release for the past through openness to a new future." J. Wentzel van Huyssteen and Erik P. Wiebe, *In Search of Self: Interdisciplinary Perspectives on Personhood* (Grand Rapids, MI: Eerdmans, 2011), p. 9.

12. Antonio Damasio tells David's story: "David was being brought to a bad-guy encounter and as he turned into the hallway and saw the bad guy awaiting him, a few feet away, he flinched, stopped for an instant, and only then allowed himself to be led gently to the examining room. I picked up on this and immediately asked him if anything was the matter, if there was anything I could do for him. But, true to form, he told me that, no, everything was all right after all, nothing came to his mind, except, perhaps, an isolated sense of emotion without a cause behind that emotion. I have no doubt that the sight of the bad guy induced a brief emotional response and a brief here-and-now feeling. However, in the absence of an appropriately related set of images that would explain to him the cause of the reaction, the effect remained isolated, disconnected, and thus unmotivated." Antonio Damasio, *The Feeling of What Happens: Body and Emotion in the Making of Consciousness* (San Diego: A Harvest Book, 1999), p. 46.

13. Damasio states, "To be sure, there was nothing in David's conscious mind that gave him an overt reason to choose the good guy correctly and reject the bad one

correctly. He did not know why he chose one or rejected the other; he just did. The nonconscious preference he manifested, however, is probably related to the emotions that were induced in him during the experiment, as well as to the nonconscious reintroduction of some part of those emotions at the time he was being tested. David had not learned new knowledge of the type that can be deployed in one's mind in the form of an image. But something stayed in his brain and that something could produce results in non-image form: in the form of actions and behavior. David's brain could generate actions commensurate with the emotional value of the original encounters, as caused by reward or lack thereof." Ibid., p. 46.

14. Jeffrey Masson, *Dogs Never Lie About Love: Reflections on the Emotional World of Dogs* (New York: Broadway Books, 1998), p. 9.

15. Morey, *Dogs,* p. 233.

16. In Stanley Coren's *The Modern Dog: How Dogs Have Changed People and Society and Improved Our Lives* (New York: Free Press, 2008), pp. 148–49 and 227. Coren continues: "The strong connection between humans and animals has been a subject of serious psychological research. Scientific evidence about the health benefits of such a relationship was first published about thirty years ago, when a psychiatrist—Aaron Katcher of the University of Pennsylvania—and a psychologist— Alan Beck of Purdue University—measured what happens physically when a person pets a friendly and familiar dog. They found that the person's blood pressure lowered, his heart rate slowed, breathing became more regular, and muscle tension relaxed—all signs of reduced stress."

8. BONDING FEVER

1. Bowlby, *Attachment,* p. xxix.
2. Lieberman discusses what I call the "pull of attachment": "We all inherited an attachment system that lasts a lifetime, which means we never get past the pain of social rejection, . . . just as we never get past the pain of hunger. We have an intense need for social connection throughout our entire lives. Staying connected to a caregiver is the number one goal of an infant. The price for our species' success at connecting to a caregiver is a lifelong need to be liked and loved, and all the social pains that we experience that go along with this need." Lieberman, *Social,* p. 48.
3. Bradshaw, *Dog Sense,* p. 68.
4. Morey gives a little background on cats in relation to dogs and attachment. He says, "It is noteworthy that a telephone survey of several hundred Rhode Island residents, reported by Albert & Bulcroft, found that '[o]wners who selected dogs as their favorite pets reported feeling more attached to their pets than did people whose favorite pets were cats or other animals.' The reasons for this kind of difference in perception probably stem from the fact that cats are solitary and mostly nocturnal hunters. In contrast, both dogs and people are socially gregarious and, in general, active by day, though wolves do sometimes hunt at night. Those are behavioral factors that surely underlie the differences in perception, and it is especially useful to focus more closely on their different patterns of sociality. In that light, it is relevant to note that in a study emphasizing visual cues, Miklosi et al. found that cats seem to lack some components of human attention-getting behavior that are present in dogs." Morey, *Dogs,* p. 204.

5. "The extraordinary similarity between the bonding processes of dogs and humans is also established by tests that have shown that the evident depression of dogs separated from their masters can be alleviated by precisely the same medications as that of humans. It is quite certain that the same biochemical processes take place in both." Csányi, *If Dogs Could Talk,* p. 55.

6. Masson, *Dogs Never Lie About Love,* p. 40.

7. Hare and Woods, *The Genius of Dogs,* p. 266. Bradshaw adds, "The remarkable thing about this strong physiological response is that it is triggered by contact with *Homo sapiens,* a different species. As noted earlier, dogs' attachment to people is often more intense than attachment to individuals of their own species; dogs that become very upset when their owners go out are rarely comforted by the presence of other dogs. It's tempting to speculate that 'one man dogs' may lack oxytocin, but so far no one has looked into this possibility. What scientists do know, however, is that all dogs have been programmed by domestication to have intense emotional reactions toward people. This lies at the root of the 'unconditional love' that many owners describe and treasure in their dogs. Such intense feelings are not easily turned off, as attested by the high proportion of dogs that hate being left alone (as many as one in five, according to one of my surveys)." Bradshaw, *Dog Sense,* p. 171.

8. See Jon Katz, *The New Work of Dogs: Tending to Life, Love, and Family* (New York: Random House, 2004).

9. Katz says, "At about the same time television and computer screens became pervasive, the emotional and familial lives of Americans also grew more complex. The extended family began to shrink and disintegrate. Di-

vorce rates shot up and remain high. The medical advances that helped people live longer also meant that the number of widowed and disabled rose sharply. Americans moved more frequently, leaving their communities. The number of single-person households continues to increase, along with the number of childless couples." Ibid., p. 13.

10. "John Archer . . . wrote in 1996 in the journal *Evolution and Human Behavior* that it's the social arrangements of modern Western societies that drive the growing bonds humans forge with pets. Mobility and divorce have gnawed away at the extended family, so family units are growing smaller. The trend toward smaller households has reached its logical and inevitable conclusion: more people live alone." Ibid., p. 116.

9. DOGGY PLAY AS SOUL TALK

1. Bradshaw, *Dog Sense,* p. 206.
2. George E. Vaillant, *Spiritual Evolution: How We Are Wired for Faith, Hope, and Love* (New York: Broadway Books, 2008), p. 129.
3. "The subject of play has turned us from individuals and their concern with identity (although this theme itself is located within the horizon of the social relations of individuals) to the life-world which individuals share. In play, human beings put into practice that being-outside-themselves to which their egocentricity destines them. The process begins with the symbolic games of children and finds its completion in worship." Ibid., p. 338.
4. Brian Sutton-Smith, *The Ambiguity of Play* (Cambridge, MA: Harvard University Press, 1997), p. 158.
5. Robert Bellah, *Religion in Human Evolution: From the*

Paleolithic to the Axial Age (Cambridge, MA: Belknap Press, 2011), p. 104.

6. Bradshaw, *Dog Sense,* p. 207.

10. A HEALING PRESENCE

1. Jonah Engel Bromwich, "In a Shaken Orlando, Comfort Dogs Arrive with 'Unconditional Love,'" *New York Times,* June 16, 2016.
2. Rachel McPherson, *Every Dog Has a Gift* (New York: Jeremy P. Tarcher/Penguin, 2010), pp. 20–21.
3. Ibid., p. 27.

11. COULD IT BE LOVE THEY'RE FEELING?

1. I believe my position here still holds, but Konrad Lorenz adds nuance by pointing out that dogs are able to understand words linked together. My point is only that they have no mastery of syntax and that they read our emotions (mind) more than the hard data of words. Lorenz adds: "It is a fallacy that dogs only understand the tone of a word and are deaf to the articulation. The well-known animal psychologist Viktor Sarris proved this indisputably with three Alsatians called Harris, Aris and Paris. On command from their master, 'Harris (Aris, Paris), Go to your basket,' the dog addressed and that one only would get up unfailingly and walk sadly but obediently to his bed. The order was carried out just as faithfully when it was issued from the next room whence an accompanying involuntary signal was out of the question. It sometimes seems to me that the word recognition of a clever dog which is firmly attached to its master extends even to whole sentences. The words, 'I must go now' would bring Tito and Stasi to their feet at once even when I exercised great self-control and spoke

without special accentuation; on the other hand, none of these words, spoken in a different connection, elicited any response from them." Konrad Lorenz, *Man Meets Dog* (London: Routledge, 1954), p. 130.

2. Berns, *How Dogs Love Us,* p. 193.

3. Ibid., p. 204. "The pattern of activations in the cortex suggested that [dogs] concocted mental models of our behavior, which might be due to mirror neuron activity. But regardless of the mechanism, the . . . data showed that their mental models included the identity of important people in their lives that persists even when the people aren't physically present." Ibid., p. 204.

4. Masson, *Dogs Never Lie About Love,* p. 210.

12. BOBBY AND THE "RIGHTEOUS" DOGS OF EGYPT

1. Emmanuel Levinas, *Difficult Freedom: Essays on Judaism* (Baltimore, MD: John Hopkins University Press, 1990), p. 153.

2. There are other biblical views and interpretations aside from Levinas's. Hobgood-Oster explains: "Dogs are equally reviled, albeit for different reasons. In the description of the holy city in the book of Revelation, dogs are left outside the gates (Revelation 22:15), thus symbolically excluded from heaven. Jesus also questions the worthiness of dogs to receive scraps from the table, though he is challenged and, some would argue corrected, by the Syrophoenician woman (Mark 7:25–30). Dogs, in this instance, are a symbol for the other but also function as real dogs who do eat scraps from human tables." Hobgood-Oster, *Holy Dogs and Asses,* p. 44.

3. Laura Hobgood-Oster, *A Dog's History of the World: Canines and the Domestication of Humans* (Waco, TX: Baylor University Press, 2014), p. 102. Aaron S. Gross

helpfully explains that this interpretation of the text as Hobgood-Oster is interpreting it is from a rabbinic interpretation: "[As] one Rabbinic interpretation of Exodus has it, [the dogs] did not bark and thereby allowed the ancient Israelites to escape from bondage." Aaron S. Gross, *The Question of the Animal and Religion: Theoretical Stakes, Practical Implications* (New York: Columbia University Press, 2014), p. 129.

13. A STORY FROM THE CAVE OF DREAMS

1. Paleoanthropologists haven't determined whether this child was a boy or a girl, but for the sake of this midrash, I am calling him a boy.
2. I'm following Mark Derr in believing that these prints are from a wolf-dog. He makes a strong case for it in his book *How the Dog Became the Dog* (New York: Overlook Duckworth, 2011). However, again, it must be mentioned that this is a debated point. In a footnote on page 82 of Morey's *Dogs,* this child and her wolf-dog are discussed further. Morey points out that it is contested that the child and wolf-dog were actually together, though I feel the evidence supports that they were.
3. It is true that some theorists believe the dog prints come years (even thousands of years) after the child left his or her footprints. This is possible, and Homans (later in this note) nicely articulates the conflicts with this theory. Yet, while this episode is contested, there are others who believe the footprints are indeed following the boy (Mark Derr, for example). I, then, am taking a leap in saying the wolf-dog is with the boy (or girl) and stand with solid thinkers who believe they are together, but the reader should be aware that this is a contested episode in the archaeological record. Homans explains the

contestation: "Chauvet Cave in southern France (which also features a gallery of animal portraits), contains a remarkably well-preserved set of child's footprints, carbon-dated at some twenty-six thousand years old. The prints are intertwined with those of a wolf, or possibly a dog, judging by the shortness of one of the digits on its front paw. So far the animal's prints have been found only on top of the child's, suggesting that the wolf—or dog—came afterward. If one of the animal's prints were to be found under the child's, it would provide a simple, elegant archaeological proof that the two had walked together. It's a beautiful image, a child and his best friend making their way through a dark place by torchlight, but thus far no such print has been found, and it remains just a dog lover's dream." Homans, *What's a Dog For?*, p. 121. Laura Hobgood-Oster, for instance, holds to the child and wolf-dog being together. She says, "Over 26,000 years ago, in the deep, damp darkness of a cave in southwestern Europe, a human child and a canine walked side by side. The child's torch brushed up against the wall numerous times, providing a time signature; and their feet left a 150-foot-long trail that marks the journey they took together. The archaeological discovery of the footprints of the young human and the wolf-dog offers a compelling snapshot of the beginnings of a story of two species whose lives would be intertwined for thousands of years to come." Hobgood-Oster, *A Dog's History of the World*, p. 5.

4. Lorenz, too, crafts a story about the meat toss: "Then suddenly the young leader with the high forehead does something remarkable and, to the others, inexplicable: he throws the carcass to the ground and begins to rip off a large piece of skin to which some flesh still adheres.

Some young members of the band, thinking that a meal is about to be distributed, come close, but with furrowed brow, the leader repulses them with a deep grunt of anger. Leaving the detached pieces on the ground, he picks up the rest of the carcass and gives the signal for marching. . . . Not a man is conscious that he has just witnessed an epoch-making episode, a stroke of genius whose meaning in world history is greater than that of the fall of Troy or the discovery of gunpowder. He acted on impulse, hardly realizing that the motive for his action was the wish to have the [wolf] near him." Lorenz, *Man Meets Dog*, p. 5.

5. John Bradshaw, in his book *Dog Sense* (see pp. 48ff.), believes that there was some form of feeding that led to domestication. This feeding Bradshaw connects to the drive to keep pets. What Bradshaw doesn't do, that I have here done drawing from Bellah, is to connect this keeping of pets with a spiritual dynamic born within *Homo sapiens*. Bradshaw, however, it must be mentioned, would not concur with my early thesis of the arrival of wolf-dog.

6. Raymond Coppinger and Lorna Coppinger, *Dogs: A New Understanding of Canine Origin, Behavior, and Evolution* (Chicago, IL: University of Chicago Press, 2002), p. 134.